JOINT INSTITUTE FOR NUCLEAR RESEARCH

LABORATORY OF NUCLEAR PROBLEMS

Venedikt P. Dzhelepov and Bruno M. Pontecorvo

INVESTIGATIONS ON HIGH ENERGY PHYSICS PERFORMED ON THE SYNCHROCYCLOTRON AT THE LABORATORY OF NUCLEAR PROBLEMS OF THE JOINT INSTITUTE FOR NUCLEAR RESEARCH

AKADEMIE-VERLAG · BERLIN

1958

Erschienen im Akademie-Verlag GmbH, Berlin W 8, Mohrenstraße 39
Lizenz-Nr. 202 · 100/542/58
Satz und Druck: Druckhaus „Maxim Gorki", Altenburg
Bestell- und Verlagsnummer: 5339 · ES 20 D 6
Printed in Germany

A survey is given of experimental investigations on high energy particles performed with the synchrocyclotron at the Laboratory of Nuclear Problems (Joint Institute for Nuclear Research) since the starting up of the accelerator to July 1957.

Introduction

Forty years since the Great October Socialist Revolution have been marked by great successes in different fields of our science and technique. Particularly fast was the development of nuclear physics.

The last fifteen years both in the Soviet Union and in the economically advanced foreign countries (the USA, England) have led to the successful solution of a number of important problems connected with practical applications of atomic energy — powerful atomic reactors and electric power stations. A new branch of technique-nuclear energetics has appeared.

The main scientific base for this branch of technique has been physical research in the low energy particle field.

This period of progress was also characterized by the birth of a new branch of physics — the physics of high energy and ultra high energy particles and powerful accelerators.

Its history dates only from the last ten years and it is not yet connected with any practical problems.

The main subject of this new field of physics is the investigation of the nature and properties of all elementary particles (nucleons, mesons, hyperons, antiparticles) which are the simplest of all the known structural elements of matter.

Various investigations performed with cosmic rays and with powerful accelerators have shown that elementary particles may be produced in collisions of particles with energies of hundreds of millions and milliards ev.

The main aim of high-energy and ultra high-energy physics is to identify the elementary particles as fully as possible, to establish the interaction laws between particles as well as to study the processes of mutual particle transformations.

If we take into account that at present twenty six different elementary particles are known and also that, in spite of the titanic work of scientists all over the world the fundamental law describing the interaction between the main structural elements of the nucleus — the nucleons — has not yet been established, then it is possible to imagine what wide and unknown horizons the new physics can present. The history of the world science shows that scientists get a lot of information not only important from a theoretical point of view but also useful for the mankind when they begin to investigate a new wide field of knowledge.

All this points to the great perspective of investigations with high energy particles.

In our paper an attempt will be made to give an idea of the main scientific investigations performed by Soviet physicists with high energy particles at the synchrocyclotron of the Joint Institute for Nuclear Research*, which is the largest accelerator of this type in the world.

STARTING UP OF THE SYNCHROCYCLOTRON AND CHARACTERISTICS OF HIGH ENERGY PARTICLE BEAMS

High energy physics research with accelerators started in our country on the 14th of December 1949 when at the Institute for Nuclear Problems of the USSR Academy of Sciences (now the Laboratory of Nuclear Problems of the Joint Institute for Nuclear Research) the five-meter synchrocyclotron was put into operation and 280 MeV deuterons and 560 MeV α-particles (1) were obtained. Practically at the same time a synchrotron accelerating electrons up to the energy of 250 MeV** began to work at the Physical Institute of the USSR Academy of Sciences.

Abroad the building of big and complicated engineering constructions such as powerful synchrocyclotrons could be achieved only by the most developed capitalist countries. (The USA, England.)

Therefore the building of the synchrocyclotron of the Joint Institute for Nuclear Research is a bright example of the high level of industrial develop-

* Since we plan to review here the results of the investigations performed with the six meter synchrocyclotron as fully as possible and since the paper is reckoned upon a wider than usual group of readers we present the material as follows: while the text is quite short, experimental data in form of diagrams are taken from original papers, with explanations containing more detail than usual. We are not concerned here with a large number of investigations made with radiochemical methods, as they will be described in another paper (A. N. Murin). Since our aim was to review the investigations performed by Soviet physicists, research made with powerful accelerators abroad are neither discussed nor quited.

** In our paper we are concerned with scientific investigations made only at the synchrocyclotron.

ment and technique of production achieved in the USSR in the years of Soviet power.

The great effort made in connection with the construction of powerful accelerators in our country illustrates the special attention and care devoted by the Communist Party, the Soviet Government and all the Soviet people to the development of advanced science.

The operation of high energy accelerators is based on the principle of phase stability in cycling resonance accelerators, which was discovered by V. I. Veksler (1944) and McMillan (1945).

The construction of our powerful synchrocyclotron has been the result of intensive activity of a large body of scientists and engineers working in a number of research institutes lasting almost three years. For a number of years these problems were in the field of attention and activity of Academician I. V. Kurchatov a fact which contributed to their successful fulfilment.

The main equipment of this powerful accelerator was built at the plants of the Ministry of Electric Industry of the USSR.

At the end of 1950 the five-meter synchrocyclotron was convented to the proton regime and accelerated protons up to about 500 MeV (1).

A considerable part of the first physical investigations was devoted to the determination of the parameters of the high energy particle beams (intensity, energy and angular distributions etc.).

In Table 1 the main results of these experiments, which were made with the accelerator working at different regimes, are given.

Table 1

	Kind of accelerated particles and their energy		
	280 MeV deuterons	560 MeV α's	480 MeV protons
Internal target current (μA)	1	0,025	0,2—0,3
Extracted proton flux at a distance of 10 m from the magnetic channel (cm^{-2} sec^{-1})	—	—	$1 \cdot 10^6$ (E_p = 460 MeV)
Neutron flux at the maximum of the angular distribution 2 m from the internal target (cm^{-2} sec^{-1})	$8 \cdot 10^7$	$2 \cdot 10^5$	$5 \cdot 10^6$
Neutron energy at the maximum of the energy distribution (MeV)	120	120	380
Halfwidth of the angular neutron distribution (radian)	0,17	0,35	0,55
Process responsible for neutron production	Stripping	α-particle disintegration	charge exchange

Nuclear investigations made on the five-meter variant of the synchrocyclotron will be reviewed in one of the following sections.

RECONSTRUCTION OF THE ACCELERATOR AND ITS PRESENT STATE

After a number of nuclear investigations with deuterons, α-particles and 500 MeV protons was carried out, in 1953 the accelerator was reconstructed, i.e., the diameter of the magnet polar expansions has been increased up to

Fig. 1. Synchrocyclotron building

six meters and the energy of the accelerated protons has been brought up to 680 MeV (1).

A view of the building where the accelerator is housed and the general view of the six-meter synchrocyclotron are given in figs. 1 and 2.*

During the reconstruction a measuring pavilion protected from radiations by massive concrete „the so called meson laboratory" and a cabin for experiments with the polarized proton beam (fig. 3) were built.

*) Details on the six-meter synchrocyclotron were given earlier in a number of special papers (1, 2, 3) and we shall not be concerned with them here.

Fig. 2. General view of the synchrocyclotron. The accelerating chamber is placed between the electromagnet poles; the high vacuum pump can be seen to the left, the high frequency generator and the frequency variator in the centre

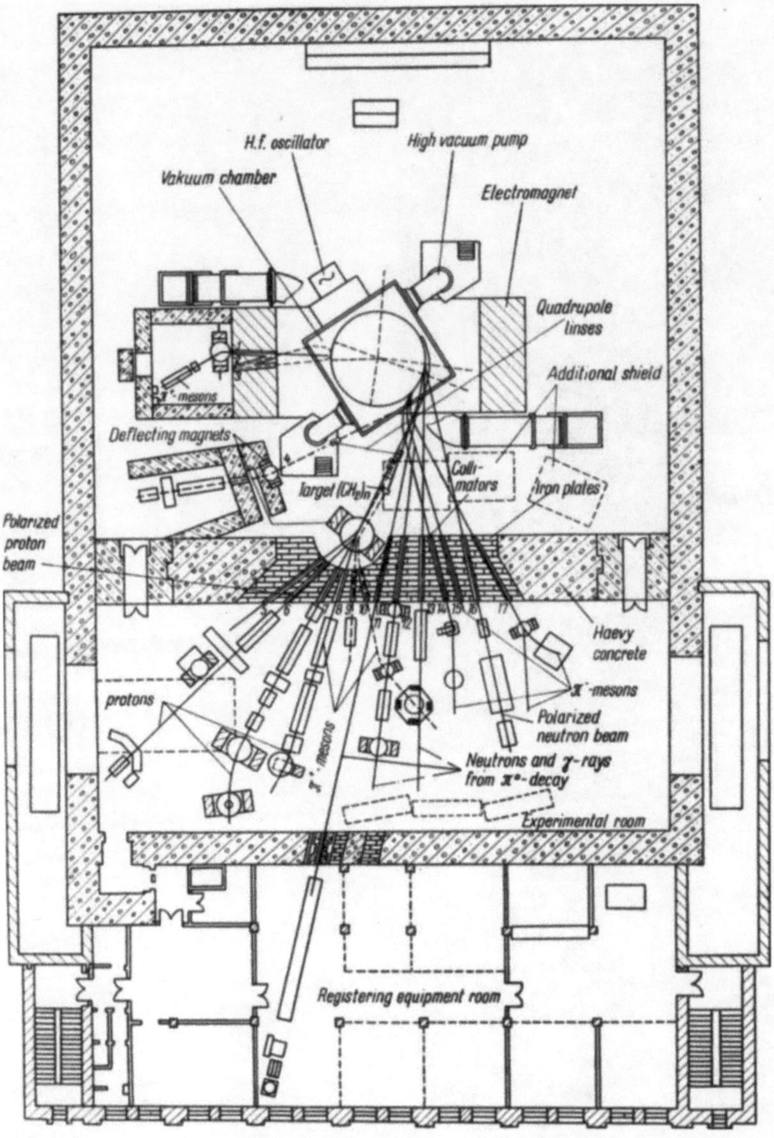

Fig. 3. Particle beams from the six-meter synchrocyclotron

In the main shield there is an embrasure for the extraction of various beams through a number of collimators.

The improvement of the conditions of particle acceleration in the central region of the chamber which has been achieved during this year made it possible to increase the proton current at the target almost two times and bring it from 0,3 μA (1) to 0,5 μA (4).

14 various beams (5,6) of particles (protons, neutrons, π^+ and μ^+-mesons, γ-quanta etc.) (fig. 3) are extracted from the accelerator's chamber into the atmosphere.

The extraction of the proton beam into the atmosphere is brought about by means of the method of radial oscillations, which has been worked out in detail at our Laboratory (7). The proton beam flux extracted into the measuring pavilion was increased additionally by focusing it with quadrupole lenses (8) and is $1{,}5 \cdot 10^9$ protons $cm^{-2}\ sec^{-1}$ at a distance of 15 m from the "outlet window" of the accelerator's chamber.

The high flux of the extracted proton beam made it possible to create some more beams of high energy particles from external targets: three π-meson beams (collimators 8,9 and 10) and a polarized proton beam (collimator 6 — fig. 3). For increasing the density of charged particle beams special focusing devices suggested at our Laboratory (9) are now widely

Table 2

Intensities of high energy particle beams from the six-meter synchrocyclotron in the measuring pavilion, as well as in the meson and polarization laboratories.

Kind of particle	Energy (MeV)	Collimator number	Beam flux $cm^{-2}\ sec^{-1}$
Protons	675 ± 6	7 as well as 6 or 8	$1{,}5 \cdot 10^9$
Polarized protons	640 ± 10	4	$4 \cdot 10^5$
	610	6	$6 \cdot 10^5$
Neutrons	for the spectrum interval $500 \leq E_n \leq 650$	11, 12, 13	$(3 \div 4) \cdot 10^4$
Polarized neutrons	for the spectrum interval $450 \leq E_n \leq 600$	16	10^4
π^+-mesons	150	8	450
	300	8	1000
	310	9	1600
	360	8	150
π^--mesons	300	1	500
	330	1	200
	370	3	70
μ^+-mesons	90	8	$20 \div 30$
μ^--mesons	25	17	60
γ-quanta from π^0-meson decay	$10 \leq E_\gamma \leq 600$	12	$3 \cdot 10^3$

used together with quadrupole lenses. They are adjusted in the gaps of the deflecting electromagnets.

In Table 2 the data on fluxes and energies of the particles in different extracted beams are given (July 1957).

At the beginning of 1957 the working time of the accelerator was increased from 100 hours (1) up to 140 hours per week.

The increase of beam intensity and of working time of the synchrocyclotron extended considerably the experimental possibilities in particular made it possible to begin the study of effects with small probability such as the meson production by mesons near the threshold.

Before reviewing the investigations made with the synchrocyclotron since its starting up we may remind the reader that some of the problems worked out have been already dealt with at the session (1955) of the USSR Academy of Sciences on peaceful uses of atomic energy (10).

SOME EXPERIMENTAL INVESTIGATIONS PERFORMED WITH 280 MeV DEUTERONS AND 560 MeV α-PARTICLES

The fact that deuterons and α-particles with energies exceeding one and a half times those obtained with American accelerators were available at our synchrocyclotron made it possible to extend in the high energy region the study of nuclear processes involving these particles, and also to perform search experiments.

In this paragraph we shall be concerned with the main investigations performed during the period when the accelerator worked in the deuteron and α-particle regime (the first half of 1950).

1. Mass Determination of Mesons Produced by Particles with Energies of about Half a Milliard ev

In 1949—50 widely discussed was the problem whether mesons of different masses exist in cosmic rays. In connection with this experiments to analyze the masses of the mesons produced by particles with an energy of about 500 MeV were made.

The experiments performed in 1951 by two groups of research workers (11,12) using different methods, have shown that 560 MeV α-particles and 490 MeV protons (13) produced only π-mesons.

The yield of the mesons of other masses, if any, was estimated (13) to be at least 200 times less than the negative π-meson yield.

2. *Nucleor Fission Induced by π-mesons*

The search for meson induced fission was accomplished by observing the fission of heavy nuclei by negative π-mesons stopped in a photoplate containing uranium and wolfram. This phenomenon was first discovered in (14, 15) and was then investigated (16).

3. *The Disintegration of 560 MeV α-particles into separate Nucleons*

When exposing internal targets made of different materials to 560 MeV α-particles intensive neutron and proton beams with mean energy of about 120 MeV and directed in a relatively narrow cone along the line of motion α-particles were found (17).

Some results of these experiments concerning the neutron flux, the angular distribution and the neutron energy are given in Table 1 (line 3).

In other experiments (18) simultaniously conducted it was established that in photoemulsions exposed to α-particles a relatively high yield of nuclear disintegrations took place which were accompanied by the emission of one or two protons in a direction close to that of the incident α-particle and with an energy equal on the average to one quarter of the energy of α-particle.

All these experimental facts make it possible to draw the conclusion that as a result of the collision between an α-particle and a nucleus the first is fully disintegrated with great probability and its nucleons are emitted in the form of a relatively narrow beam. The total neutron yield of the 560 MeV α-particle break-up process is not much less than the yield from the ,,stripping process" of 280 MeV deuterons (Table 1).

4. *Deuteron "stripping"*

This phenomenon has been discovered at Berkeley and was investigated there for the deuteron energies of 190 MeV. In our laboratory the phenomenon was studied (19) with different targets at the energy of 280 MeV. Some experimental data are given in Table 1 (line 2). It was shown that angular distribution of the neutrons produced in this process as well as the total cross-section of the reaction are in satisfactory quantitative agreement with the calculations made on the basis of the ideas of Serber.

5. *Hard γ-radiation from the Synchrocyclotron Target*

When bombarding the internal target of the accelerator by 560 MeV α-particles it was discovered (20) that hard gamma-radiation is emitted the origin of which is probably due to the decay of π^0-mesons produced by α-particles.

6. *Fission of Light Nuclei by Neutrons*

In experiments with 120 MeV neutrons it was discovered (21) that nuclei of elements of the middle part of the periodic system such as dysprosium, erbium and rhodium undergo fission under neutron bombardment. It may be noted here that at about the same time the fission of silver was observed with the help of radiochemical method when bombarding it with 560 MeV α-particles. According to some theoretical concepts (23) the fission of nuclei of this part of the periodic system must occur mainly from excited states and therefore the experiments were the first proof of the existence of such a mechanism of nuclear fission.

The fission of nuclei of heavier elements (with $Z \geq 73$) under high energy neutron bombardment was investigated in a number of publications (21, 24).

7. *Artificial α-radioactivity*

The production of α-active nuclei was induced (25) by bombarding targets of different materials with 560 MeV α-particles. As a result of the investigations there was found a number of new α-active nuclei of Gd, Tb and Dy, which were produced by bombarding Ho, Yb, Tb and Er. Among several other investigations we may mention here the study of the disintegrations of nuclei by π-mesons (26) and high energy deuterons (27). Inelastic interaction of high energy neutrons with nuclei (28) has been studied too. Some work was devoted (29) to the investigation of the emission of secondary neutrons by nuclei from Be to Pb under the effect of neutrons with the mean energies of 120 and 380 MeV.

ELASTIC NUCLEON-NUCLEON SCATTERING

One of the central problems in modern nuclear physics is that of nuclear forces, i.e., forces acting between the nucleons.

Experiments on nucleon scattering performed with low energy particles made it possible to establish that nuclear forces are not electromagnetic (they differ from those by their high intensity and their very small radius of action) and are dependent on the spins of the particles in a very complicated way.

A number of other experimental facts pointed out also to one important property of these forces — their symmetry (the approximate equality of nuclear interactions between two protons and between two neutrons) — and made it possible to suggest the fargoing hypothesis of charge independence of nuclear forces. This hypothesis postulates the equality of nuclear interactions between any two pairs of nucleons in states with equal total spins and orbital moments.

New data on nuclear forces may be obtained from experiments on nucleon-nucleon scattering of particle energy of hundreds MeV when the wave length of the incident particle becomes essentially less than the range of nuclear forces ($\lambda < \hbar/\mu c = 1{,}4 \cdot 10^{-13}$ cm).

In order to clear up the extent to which forces dependent on spin influence the scattering it is necessary to make experiments with non-polarized particle beams as well as with polarized nucleon beams.

It is worth pointing out that methods used for the theoretical analysis of experimental material used at low energies turn out to be quite unfit in the new energy region. Because of this and due to the absence of a strict theory of nuclear forces the consequences of the hypothesis of charge independence of nuclear forces turned out to be very fruitful. The use of the isotopic spin apparatus makes it possible to go forward in the analysis of the results of the experiments with high energy particles and to find out new regularities in the behaviour of nuclear interactions (30).

This apparatus makes it possible to break up the complex of possible states in which the system of two nucleons at high collision energy may be present into two groups with the values of total isotopic spin $T = 1$ and $T = 0$ and to investigate the interaction in these groups of states separately.

The data on the interaction of two nucleons in the states with $T = 1$ may be obtained investigating the scattering of any nucleon pair ($p-p$, $n-n$, $n-p$), while the investigation of nucleon interaction in states with $T = 0$ requires experiments on the scattering of different nucleons ($n-p$).

In view of the great importance of nucleon-nucleon scattering investigations they have been conducted at the synchrocyclotron of the Laboratory of Nuclear Problems on a large scale since the fall of 1950. The experiments have been performed with nucleons having the energy from 380 up to 660 MeV, i.e., in the energy region which had not yet been investigated at the time.

1. Elastic Proton-Proton Scattering and Polarization in elastic p—p Scattering

In the investigations made by Segrè, Chamberlain and others (the USA) in 1949 there was established that in the energy region 150—340 MeV the cross-section of elastic proton-proton scattering does not depend upon the angle and upon the energy.

Proton-proton scattering has been studied at the synchrocyclotron of the Laboratory of Nuclear Problems since 1952 (31—33), the proton energy being from 460 up to 660 MeV.

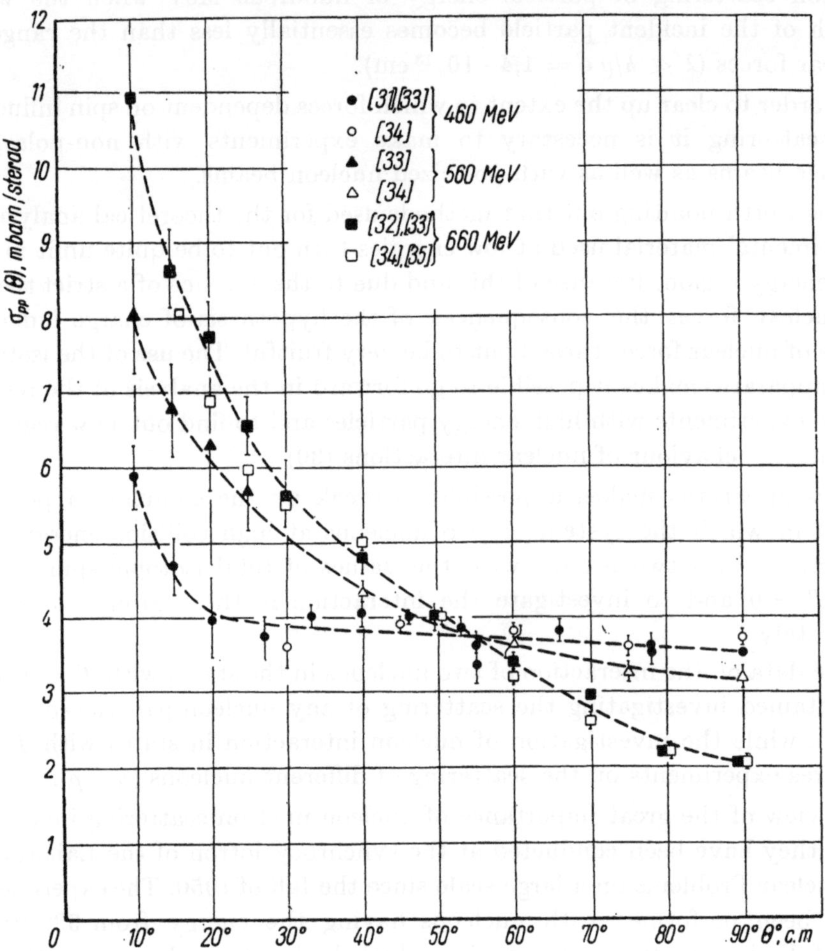

Fig. 4. Proton-proton elestic scattering at 460, 560 and 660 MeV

In these researches there was first found sharp anisotropy in high energy proton-proton scattering, the differential cross-section increasing with decreasing scattering angle.

Analogous results have been obtained by a second group of research workers (34, 35), who used another experimental method.

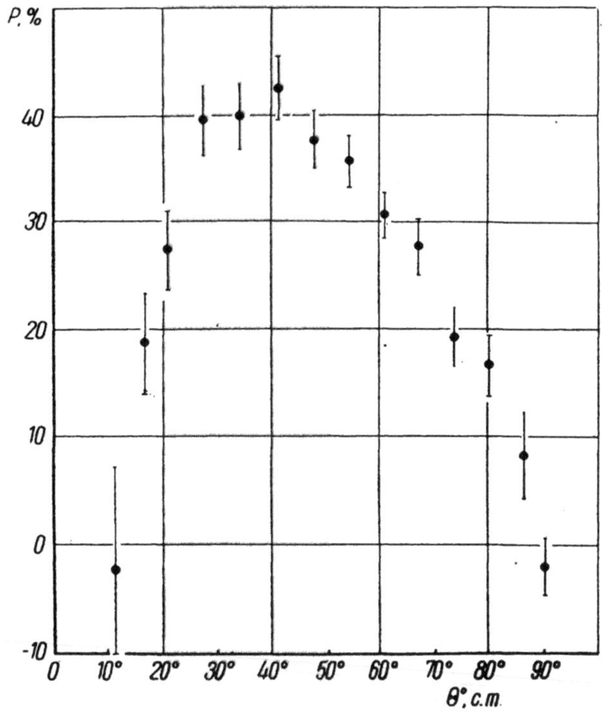

Fig. 5. The dependence of the polarization $P(\Theta)$ on scattering angle in elastic proton scattering at 635 MeV. A definite contribution from triple F-states is apparent from an analysis of the data

The experimental data obtained (31—35) for three energies of the incident protons are given in Fig. 4.

Together with the anisotropy constancy of total elastic proton-proton cross-sections which was found (31—33) in the investigated energy interval is worth noting.

Phase shift analysis of experimental data on elastic nucleon-nucleon scattering requires information on polarized particle scattering besides information on non-polarized nucleon-nucleon scattering.

The results of the investigation of polarization effects in elastic proton-proton scattering (36) at 635 MeV are given in Fig. 5.

2. *Elastic neutron-neutron scattering*

In the investigations (37, 38) made at the Laboratory of Nuclear Problems in 1952 and 1955—56 data on neutron-neutron scattering have been first obtained by means of original experimental methods. Differential cross-sections of elastic neutron-neutron scattering were measured at neutron energies of 300 and 590 MeV.

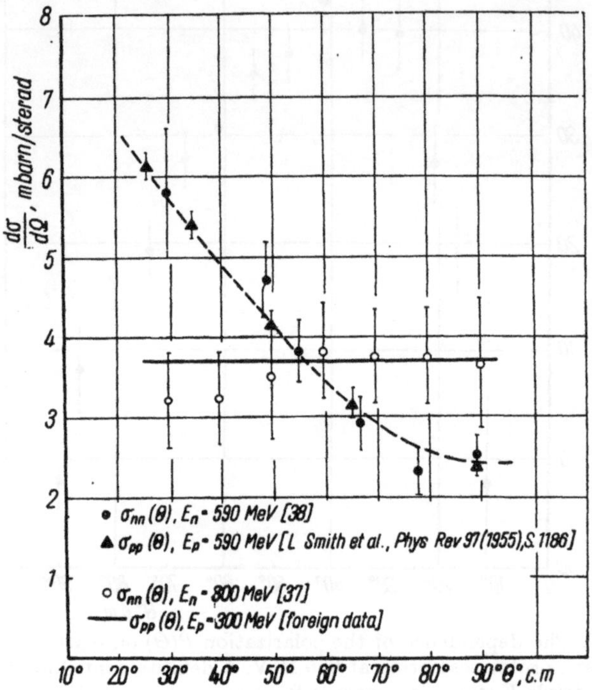

Fig. 6. Neutron-neutron elastic scattering at 300 and 590 MeV

Within the limits of experimental accuracy differential cros-sections in neutron-neutron and proton-proton elastic scattering at similar nucleon energies were found to be equal in all the region of scattering angles (fig. 6).

Total $(n-n)$ and $p-p$ interaction cross-sections were found to be equal too (39, 40). This most directly confirms that the hypothesis of charge symmetry of nuclear forces in the high energy region is correct and makes it possible to consider all the main conclusions concerning nuclear interaction of two protons to be correct for the interaction of two neutrons.

3. *Elastic Neutron-Proton Scattering*

Experiments on neutron-proton scattering are important because they yield data on the nuclear interaction between non-similar nucleons. The first experiments on high energy neutron-proton scattering were made by Segrè and others with 40 and 90 MeV neutrons in 1948 at Berkeley; an intensive backward neutron scattering (in the centre of mass system) gave a direct proof that the forces acting between neutron and proton are exchange forces.

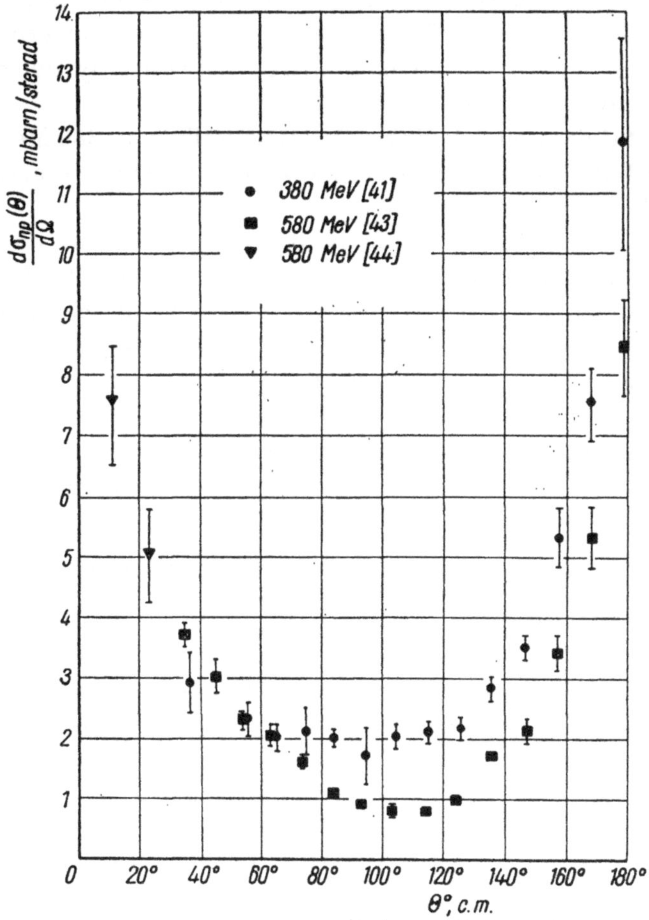

Fig. 7. Neutron-proton elastic scattering at 380 and 580 MeV. An analysis of the data at 580 MeV shows that states with orbital moment up to $l \sim 5$ are present. The sharp increase of the n—p scattering cross-sections in the region of small angles at 580 MeV is due, apparently, to the increase of the probability of nonelastic processes (meson production) in the energy interval of 380—590 MeV

Neutron-proton scattering has been systematically studied with the synchrocyclotron of the Laboratory for Nuclear Problems in the neutron energy region from 380 up to 580 MeV (41—44).

The results of these investigations are given in fig. 7.

The experimental data on 380 MeV neutrons obtained in 1950—51 show that the differential cross-section of $(n-p)$ scattering does not practically change in a wide region of angles near 90° and has a relatively great value.

This means that at high energies the intensity of the interaction in the neutron-proton system is very high as well as it is in the proton-proton system. Thus, on the basis of experiments on nucleon scattering there was established that interaction of high energy between any nucleon pairs is characterized by a very high intensity.

Experimental data on 580 MeV (43, 44) as well as on 380 MeV show that all the contribution of exchange forces to the nucleon interaction remains high up to the energy of 600 MeV.

It is worth noting that all the body of experimental data on elastic high energy nucleon-nucleon scattering does not contradict the hypothesis of charge independence of nuclear forces. The results of the analysis of the data on nucleon scattering made (30, 42) on the basis of this hypothesis point to a sharp difference in the character of the change with angle and energy of the cross-sections for nucleon interaction in states with different isotopic spins ($T = 0$ and $T = 1$) and consequenty indicates difference in the character of the nucleon-nucleon interaction in these states.

4. Exchange Scattering of Neutrons by Deuterons of 380 MeV

The results of all the experiments on unpolarized neutron-proton scattering do not yield information on the dependence of these forces on the spin, they point out only to the role of exchange nuclear forces. An original way for clearing up this point was suggested (45) in which it was proposed to observe the comparative number of high energy protons form exchange collisions of high energy neutron with deuterons and free protons.

In experiments of such a kind (46) there was found out, that in the region of small scattering angles the cross-section for high energy proton emission in exchange neutron-deuteron scattering is considerably less than that observed in neutron scattering on free protons (fig. 8). This fact means that in exchange $(n-p)$—collisions spin exchange occurs with considerable probability or, in other words, that the contributions to the exchange neutron interaction with proton by spin-exchange and spin-non-exchange forces are of the same order.

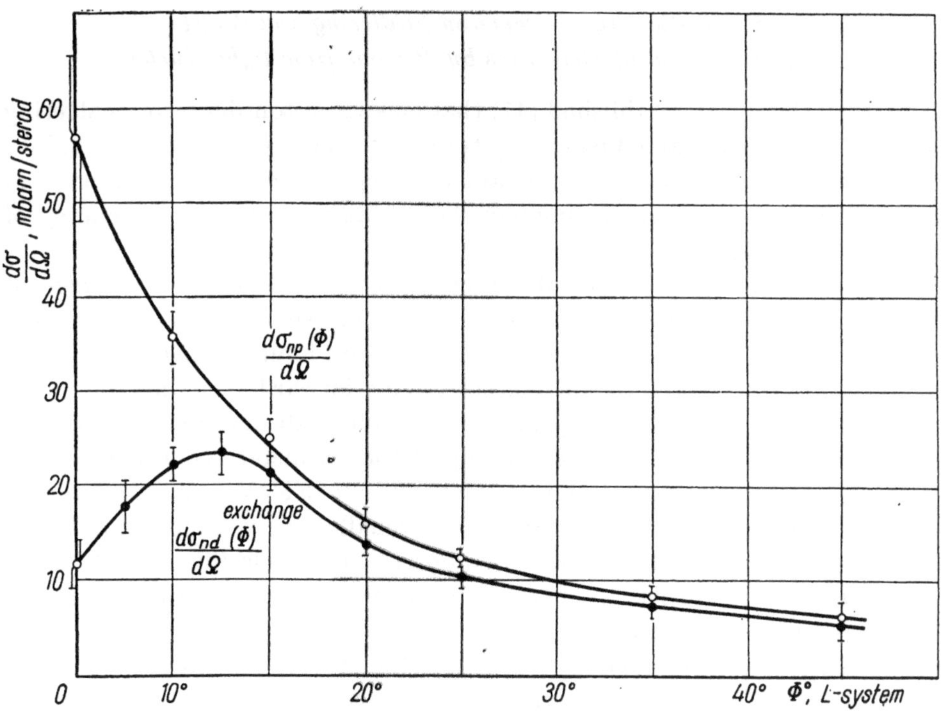

Fig. 8. Charge exchange neutron scattering by protons and deuterons at 380 MeV

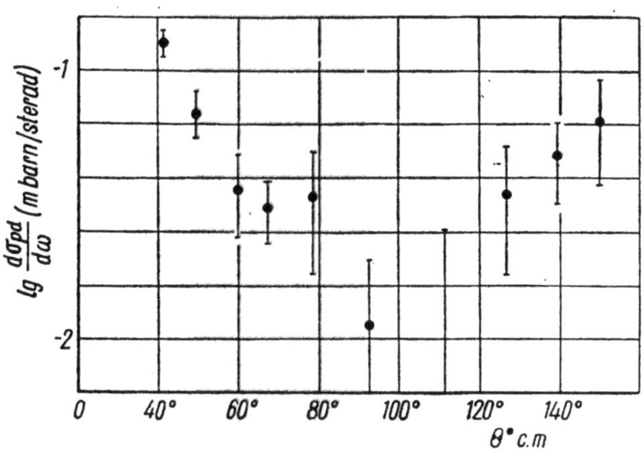

Fig. 9. Elastic proton-deuteron scattering at 660 MeV.
The total cross-section of elastic (p—d) scattering at this energy is about 2—3% of the total cross-section of (p—d) interaction

5. *Elastic Proton-Deuteron Scattering and direct Knocking-out of Deuterons by Protons from light Nuclei*

In 1952 there was established (47) that elastic proton-deuteron scattering of 460 MeV is accompanied by energy transfer to the deuteron up to 300 MeV. The most complete data on the angular dependence of the differential cross-section of $(p-d)$ scattering have been obtained for a proton energy of 660 MeV (48) (fig. 9).

Proton scattering at angles of the order of $130°-150°$ (in the centre of mass system) found in these experiments is of particular interest, when energies of $500-600$ MeV, exceeding the deuteron binding energy hundreds of times, are transferred to the deuteron flying forward as a whole.

In order to study high energy proton collisions with quasideuteron groups in nuclei special experiments were set up at the six-meter synchrocyclotron.

When bombarding nuclei of light elements by 675 MeV protons deuterons emitted with energies of about 600 MeV at small angle were found (49) just at the same energy of the particles and observation angle (fig. 10), at which such effect occured in elastic $(p-d)$ collisions which were investigated in the same experimental set up.

We can give a qualitative explanation of the observed phenomenon on the basis of the ideas about the existence of „short-life" strongly interacting nucleon groups (50) or if we use the predictions of the nuclear model lately developed by Bruekner.

All these experiments together with those — for instance, (51) — in which under the effect of high energy particles the emission of heavy fragments (T, He, Li, Be etc.) with energies exceeding the potential barrier has been observed, indicate a way for investigating the character of the forces acting between nucleons in nuclei.

6. *Total Interaction Cross-Sections of Nucleons with Nucleons and Deuterons*

The results of the main experiments set up at the six-meter synchrocyclotron by different research workers (33, 39, 42, 52) to measure total cross-sections of nuclear interactions and elastic nucleon scattering are given in fig. 11. From the results it was possible to draw the following conclusions:

1. In the energy region of $380-660$ MeV a sharp increase of the cross-sections for meson production in the collisions of both identical and different nucleons is observed.

2. Total cross-sections of elastic nucleon-nucleon scattering in states with different isotopic spins, just as at less energies, are found to have different dependence on energy: at $T = 1$ remains constant and at $T = 0$ decreases

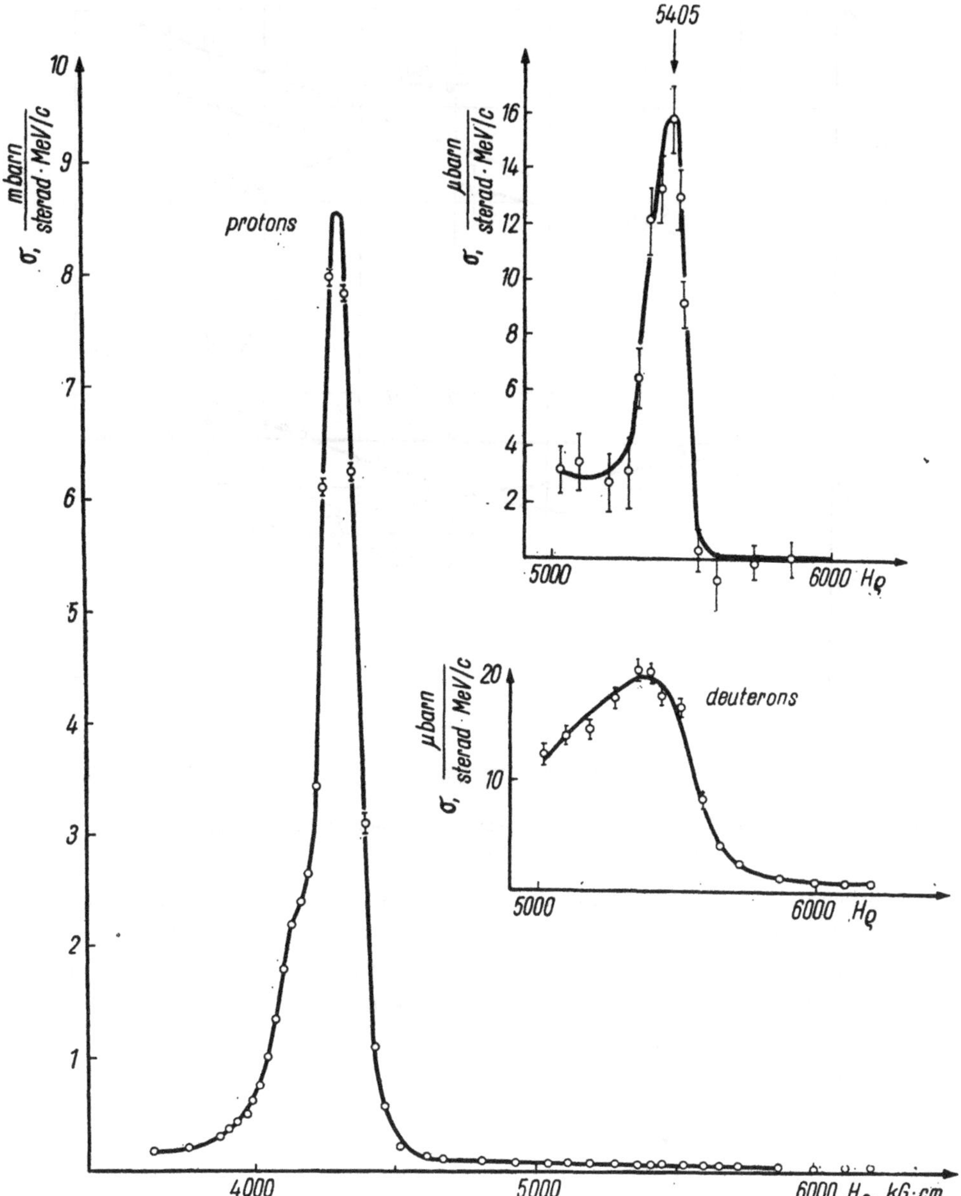

Fig. 10. Impulse spectrum of protons and deuterons emitted from lithium at 7,6° with respect to the 675 MeV proton beam. The deuterons are emitted as a result of the collisions of incident protons with quasideuteron groups inside the nuclei of lithium. The impulse deuteron spectrum from elastic (p—d)-collisions for the same angle of observation and the same energy of incident protons is given to the right at the top

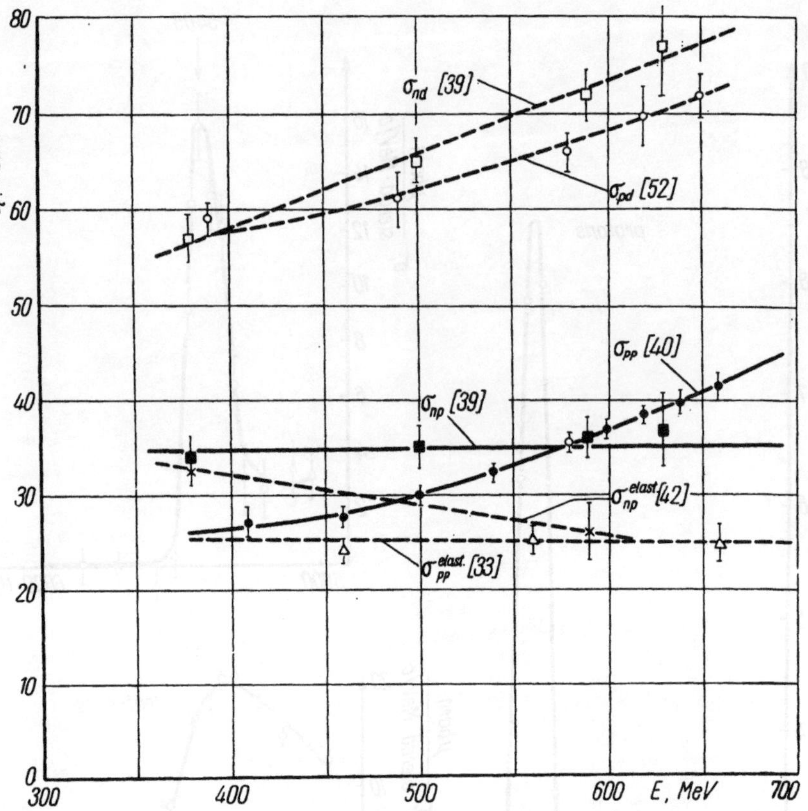

Fig. 11. Energy dependence of total cross-sections of nucleon interaction with nucleons and deuterons as well as of total cross-sections of elastic (n—p) and (p—p) scattering

with energy (42). This may point out the weakening with increasing energy of the intensity of the nucleon interaction in the states with $T = 0$.

Summarizing the results of the experimental investigations on nucleon elastic scattering by nucleons and deuterons one can state that these experiments give an idea of the main properties of nucleon-nucleon interaction in the energy region up to 0,7 BeV.

1. Very intensive forces of different types — central, tensor, exchange, dependent on spin in a complicated way contribute to the nucleon interaction at high energies.

2. Contributions to nucleon interaction from forces of different types are comparable.

3. The nucleon interaction is charge independent.

No doubt that these experimental facts have to be taken into account by the future theory of nuclear forces.

INTERACTION OF MESONS AND NUCLEONS

The problem of the nuclear forces cannot be fully separated from the study of the properties of all other elementary particles. In fact it is known that there is a connection between elementary particles and fields of forces. For instance, the interaction between charged particles is brought about by protons, in other words, the protons are the quanta of the electromagnetic field. Therefore, the properties of the photons are closely connected with the character of electromagnetic forces acting between charged particles.

By analogy the properties of π-mesons, as is now established, are closely connected with the forces acting between nucleons i. e., with the nuclear forces. Therefore, one can easily understand what an important place in elementary particle physics belongs to the investigation of π-meson properties.

Meson theory which is the formal theory of meson fields is based namely on the assumption of Yukava that nuclear forces are due to mesons. This idea is, no doubt, correct, but, unfortunately, meson theory is now at a relatively early stage of development.

Now when any concrete form of meson theory turns out to be uncapable to give anything, except, in the best case, qualitative agreement with the experiments the main problem of the experimental physics in the high energy region is collecting data necessary for a phenomenological description of the mutual interaction of different particles. As a matter of fact, as practice has shown, without a general picture uniting different phenomena of elementary particle physics it is impossible to construct a consistent meson theory. Thus, the phenomenological description of the interaction of various particles and first of all of the interaction of π-mesons with nucleons is the first step for the construction of the meson theory of nuclear forces.

Both in the USSR and abroad the investigations on the basis of which a phenomenological description of particle interaction could be obtained have been conducted in many directions. Besides, the investigation of nucleon-nucleon scattering, the meson-nucleon scattering as well as the meson production on nucleons by nucleons, mesons and γ-quanta have been studied at different energies of incident particles and scattering angles.

1. Scattering of π-Mesons on Nucleons and on Complex Nuclei

In the period 1950—54 great progress has been made in the investigation of π-meson properties thanks to the classical measurements of their parity and spin made by Panofsky and Steinberger as well as to the measurements of π-meson-nucleon total and differential cross-sections in the energy region up to 200 MeV made by Fermi and others. The investigations of meson-nucleon scattering began at the Laboratory of Nuclear Problems in 1954, after well collimated π-meson beams (6) had been obtained. The investigations were aimed at obtaining information in the meson energy region above 200 MeV where data were practically absent.

Some of the investigations were devoted to the study of the dependence on energy of the total cross-sections of π^+ and π^--mesons with hydrogen and deuterium (53, 54, 55, 56) (fig. 12) in the energy region of 140 – 400 MeV and of the angular distribution of mesons scattered on hydrogen in the reactions

$$\pi^+ + p \rightarrow \pi^+ + p, \quad \pi^- + p \rightarrow \pi^- + p, \quad \pi^- + p \rightarrow \pi^0 + n$$

at different meson energies 175, 200, 240, 270[56], 307[57-61], 330 and 360 MeV[63].

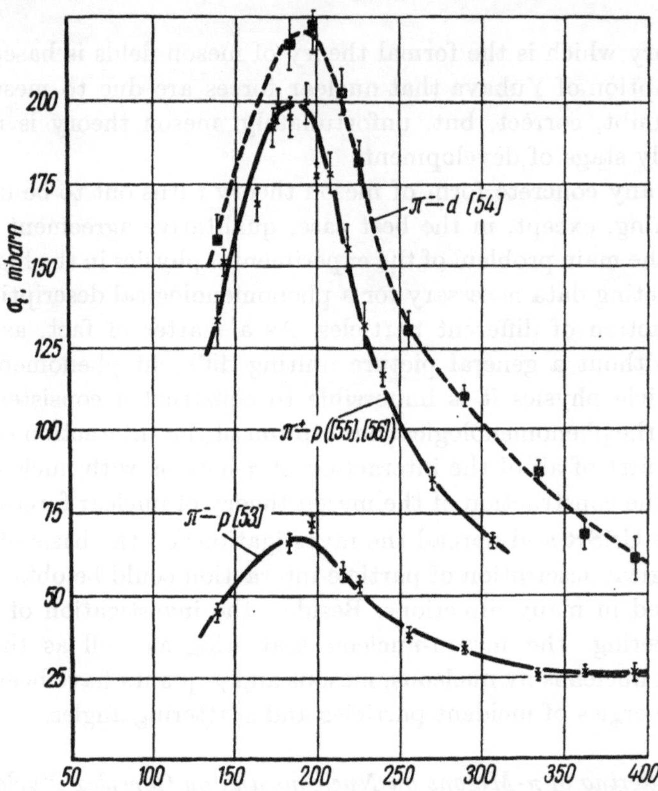

Fig. 12. Total cross-sections of π^+ and π^--mesons with hydrogen and deuterium. The „resonance" behaviour of the cross-section near 190 MeV characterized the meson-nucleon interaction in the state with isotopic spin $^3/_2$ and total angular momentum $^3/_2$. Only at energy $E_{lab} \geq 300$ MeV the contribution to the scattering of the isotopic spin $^1/_2$ state becomes noticeable

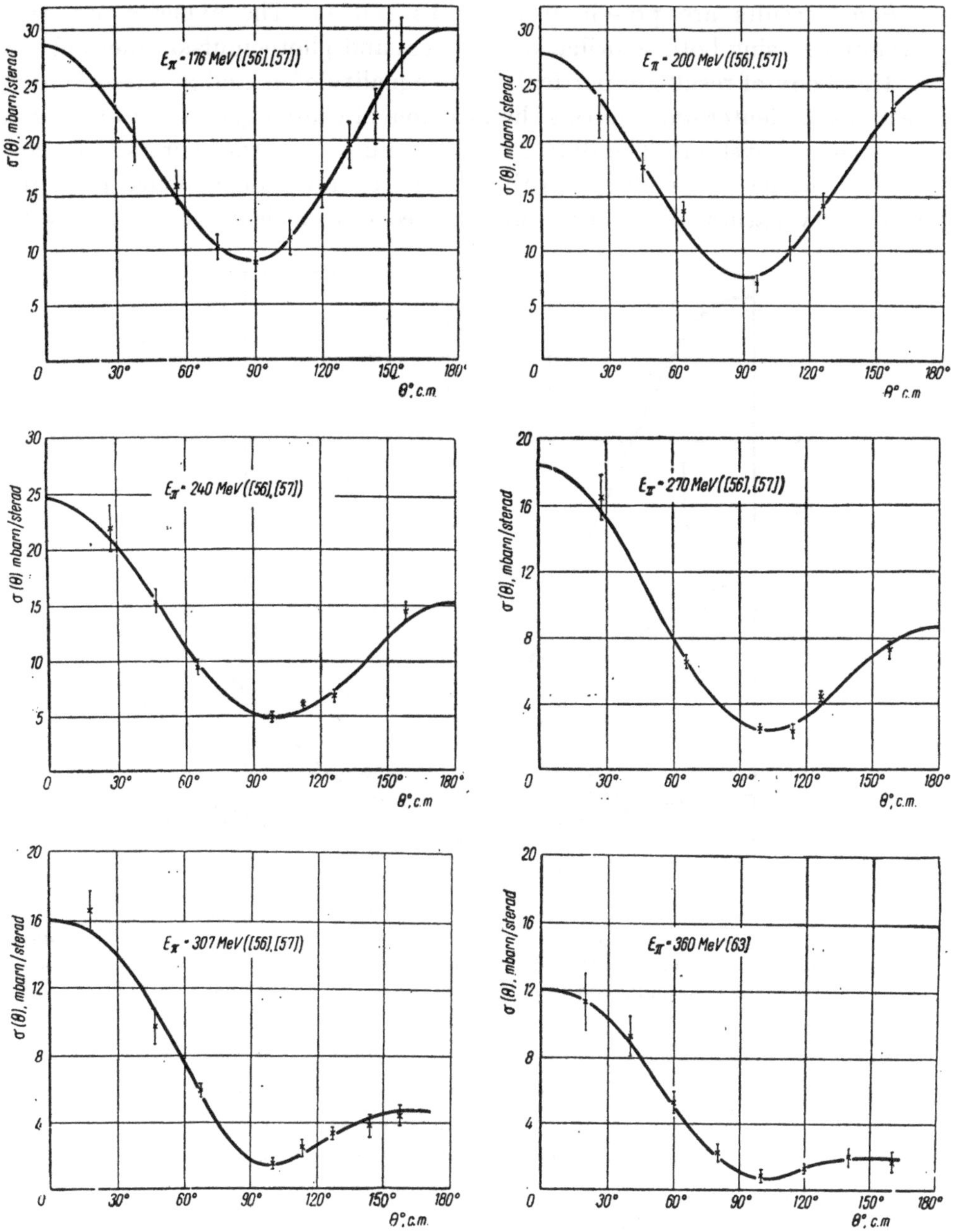

Fig. 13. Angular distribution of π^+-mesons elastically scattered on hydrogen at different energies. It can be seen from the figure that the angular distribution is symmetrical with respect to 90° near the „resonance" energy (190 MeV). At higher energies forward scattering is predominant

Some results are presented in fig. 13 and 14. The experiments were performed using both scintillation counters and photoemulsion method.

The body of results in particular, the equality of π^+ and π^--meson cross-sections on deuterium confirms the principle of charge symmetry for meson and nucleon systems. The data strongly confirm also the more strict principle of charge independence. Although the charge independence principle has a very natural place in the meson theory of nuclear forces, it may be

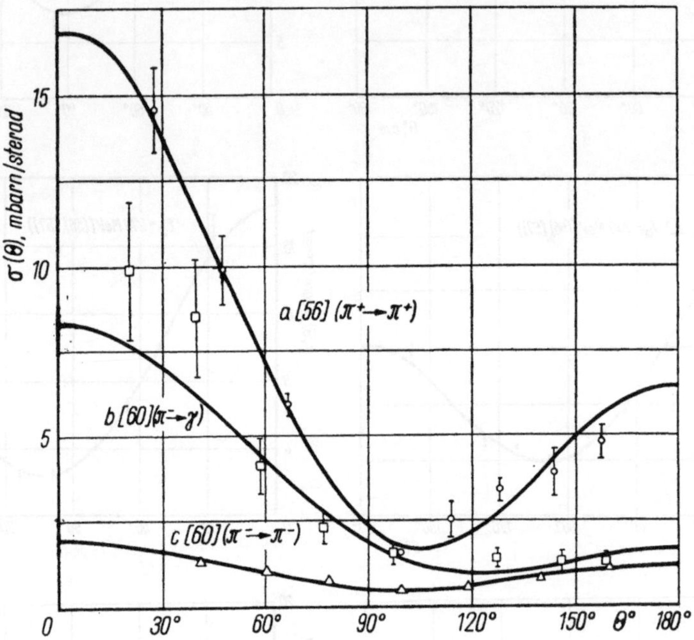

Fig. 14. Scattering of π-mesons on hydrogen at 307 MeV in the processes: a) $\pi^+ + p \to \pi^+ + p$, b) $\pi^- + p \to \pi^0 + n \to \gamma + n$, c) $\pi^- + p \to \pi^- + p$. On the basis of the data given in this figure as well as in fig. 13 one can obtain the magnitude of the meson-nucleon interaction coupling constant f^2, which turns out to be equal to $\sim 0{,}1$

noted here that the conclusions on the correctness of this principle were made without using meson theory, but only on the basis of a phenomenological analysis of the data on meson-nucleon scattering and also of the data of meson-nucleon production which will be reviewed below.

The experiments confirm that in the energy region up to 300 MeV the meson-nucleon interaction is very strong in the state with isotopic spin $^3/_2$ and total angular momentum $^3/_2$, the value of the scattering cross-section in this state being maximum at π-meson energy of about 190 MeV. This fact is expressed sometimes by the statement that the meson-nucleon interaction has resonance character. It is possible, but by no means certain, that the resonance is connected with the nucleon structure (64).

The investigation of the meson-nucleon interaction in the high energy region made it possible to conclude (53, 60) that the contribution to the scattering of the state with isotopic spin $^1/_2$ becomes notable at the energy $E_\pi \geq 300$ MeV.

The high accuracy of the angular distribution measurements of π^+-mesons scattered by hydrogen at the energy more than 200 MeV made it possible to make a phase shift analysis taking into account not only S- and P-states but also D-states. From the analysis (57) of the data it follows that the meson-nucleon interaction radius is about $7 \cdot 10^{-14}$ cm.

All the data obtained at the synchrocyclotron of the Laboratory of Nuclear Problems (56, 57, 59, 60, 62, 63) are compatible with one value of the so called meson-nucleon coupling constant f^2 which turns out to be approximately equal to 0,1.

To the meson interaction with complex nuclei was devoted a number of investigations making use of the most different experimental methods: scintillation counters (65—67), photoplates (68, 69) cloudchamber (70), diffusion chamber (71, 72). Some of results are shown in fig. 15, 16, 17 and we shall not discuss them further. It may be mentioned here that a study of the π-meson inelastic scattering in photoemulsion nuclei has shown (69)

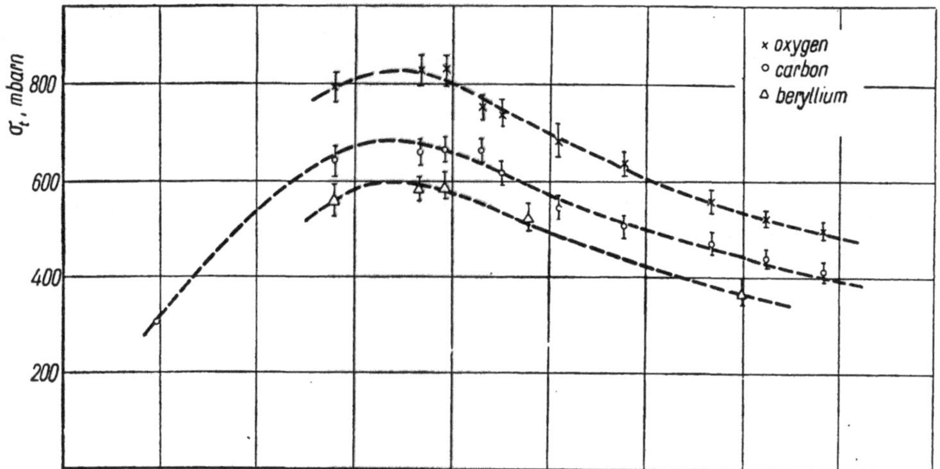

Fig. 15. The energy dependence of total cross-sections of π-meson interaction with light atomic nuclei (65). The curves remind the energy dependence of total cross-section of π^+ and π^--meson interaction with nucleons. The analysis shows (66) that pion interaction in nuclei occurs, mainly, with separate nucleons of nuclei

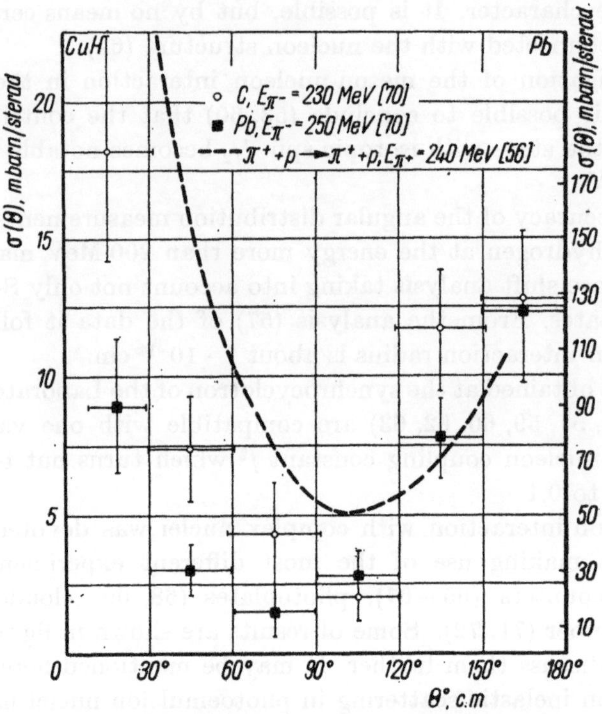

Fig. 16. Inelastic scattering of π^--mesons at 250 and 230 MeV on carbon and lead nuclei. Analysis of the results show that there exists a correlation between the energy loss and the pion scattering angle. This shows that inelastic pion scattering in the angular interval from 60° up to 180° takes place mainly as a result of scattering by separate nucleons in nuclei. The dotted curve gives for comparison the angular dependence of 240 MeV π^+-mesons elastically scattered by hydrogen

that the forces acting between nuclei and π-mesons with the energy of 160 MeV are attractive, the depth of the corresponding potential hole being approximately 25 MeV.

2. Meson Production by Nucleons

a) Early experiments.

Numerous investigations at the synchrocyclotron have been devoted to the problem of single production of charge and neutral π-mesons in nucleon-nucleon collisions. It is clear that meson production processes in nucleon-nucleon collisions are more complicated from a theoretical standpoint than meson-nucleon scattering processes.

In order to get a full enough picture of the meson production phenomena in nucleon-nucleon collisions, qualitative and quantitative characteristics of these processes were studied by using different methods.

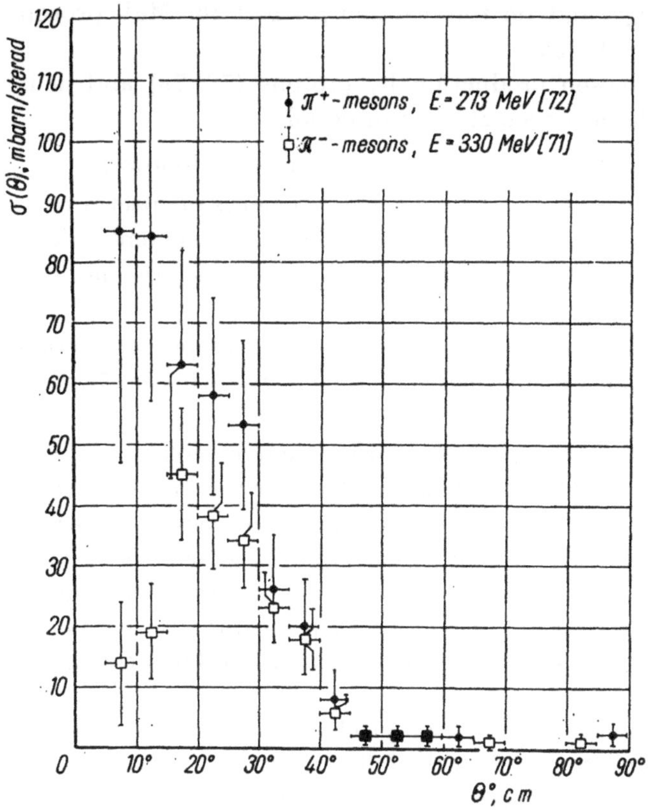

Fig. 17. Elastic scattering of π^+-and π^-- mesons by helium nuclei. The different behaviour of the scattering cross-sections at small angles is connected with the difference in sign of the Coulomb and nuclear interaction interference for π^+- and π^--mesons

In collisions between two nucleons π-mesons may de produced in the following processes:

$$p + p \to \pi^0 + p + p, \qquad n + n \to \pi^0 + n + n,$$
$$p + p \to \pi^+ + p + n, \qquad n + n \to \pi^- + p + n,$$
$$p + p \to \pi^+ + d; \qquad n + n \to \pi^- + d;$$

$$n + p \to \pi^+ + n + n,$$
$$n + p \to \pi^- + p + p,$$
$$n + p \to \pi^0 + n + p,$$
$$n + p \to \pi^0 + d.$$

Making use of the five-meter variant of the synchrocyclotron data on π^0-meson production in $(n-p)^{(73)}$ and $(n-n)^{(73)}$ collisions at 400 MeV neutron energy and on the reaction $p-p$ at 460 MeV proton energy were obtained in 1951—52.

Besides π^0-meson production by protons (78, 79) and neutrons (80) on complex nuclei has been studied in detail.

The main content of these investigations is as follows:

1. From accurate and difficult measurements of spectra of γ-quanta from the decay of π^0-mesons produced on complex nuclei bombarded with 460 MeV protons the conclusion was drawn that mesons are mainly producted in P-states.

2. Measurements of the probability of π^0-meson production in nuclei of different atomic weight made it possible to estimate (73, 74) the π^0-meson mean free path in nuclear matter.

3. Experiments on π^0-meson production in collisions of neutrons with nucleons and complex nuclei were set up for the first time (81).

4. There was shown (73) that near the meson generation threshold π^0-meson production in neutron-neutron collisions is forbidden, just as it is forbidden the production of such mesons in proton-proton collisions. This similarity of $(n-n)$ and $(p-p)$ interactions directly points out the correctness of the charge symmetry hypothesis for systems consisting of nucleons and mesons.

5. Experiments on π^0-meson production in $(n-p)$ and $(n-n)$ collisions made it possible to give (73) the following later widely confirmed qualitative scheme of π-meson generation near the meson production threshold.

Probable processes	Processes with relatively small probability.
$p + p \rightarrow \pi^+,$	$p + p \rightarrow \pi^0,$
$n + n \rightarrow \pi^-,$	$n + n \rightarrow \pi^0,$
$n + p \rightarrow \pi^0,$	$p + n \rightarrow {}^{\pi^+}_{\pi^-}.$

It was first pointed out (82) that the small probability of charged meson production in $(n-p)$ collisions is due to the fact that the meson-nucleon interaction is particularly strong in the state with isotopic spin and angular momentum $^3/_2$, i. e., in the state which was clearly recognized in the experiments on π-meson nucleon interaction.

6. Experiments in which hydrogen was bombarded by 460 MeV protons proved (74—76), and this is in agreement with the independent result of Marshall in Chicago, that with increasing energy the selection rule forbidding the reaction $p + p \rightarrow \pi^0 + p + p$ at low energies is removed. An analogous conclusion for the π^0-meson production in $(n-n)$ collisions

was obtained (83), when a 580 MeV neutron beam was obtained after the reconstruction of the synchrocyclotron.

7. The reaction $p + p \to \pi^+ + d$ was investigated (77) in detail with 560 MeV protons. The angular distribution which was obtained suggests that mesons in this reaction are produced in P-states and that D-state contribution is negligible.

Energetic Dependence of Meson Production Cross-Sections and angular Distribution of Mesons Produced by Nucleons in different Reactions

After the reconstruction of the synchrocyclotron, when a 680 MeV proton beam became available, all the processes of meson production numerated above were investigated. Since the limits of the present paper do not allow

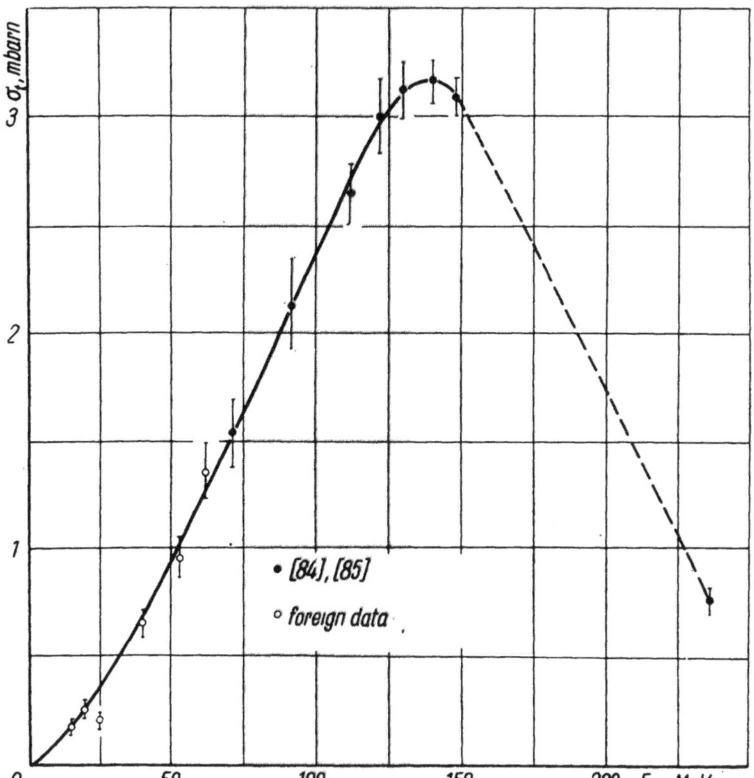

Fig. 18. The energy dependence of the total cross-section for the reaction $p + p \to \pi^+ + d$; E_π is the kinetic energy of the produced meson in the centre of mass system. The maximum is connected with the strong („resonance") meson-nucleon interaction in the state with isotopic spin $^3/_2$ and total angular momentum $^3/_2$

to give a full survey of the experiments we rely to a considerable extent upon the figures.

The study (84) of π^+-meson production in proton-proton collisions in the energy interval 460 – 660 MeV made it possible to establish for the first time that dependence upon energy of the total cross section of the reaction $p + p \rightarrow \pi^+ + d$ has resonance character (fig. 8). It follows from the analysis of angular distributions of mesons emitted in the above mentioned reaction (fig. 19) that in the considered energy region mesons are mainly

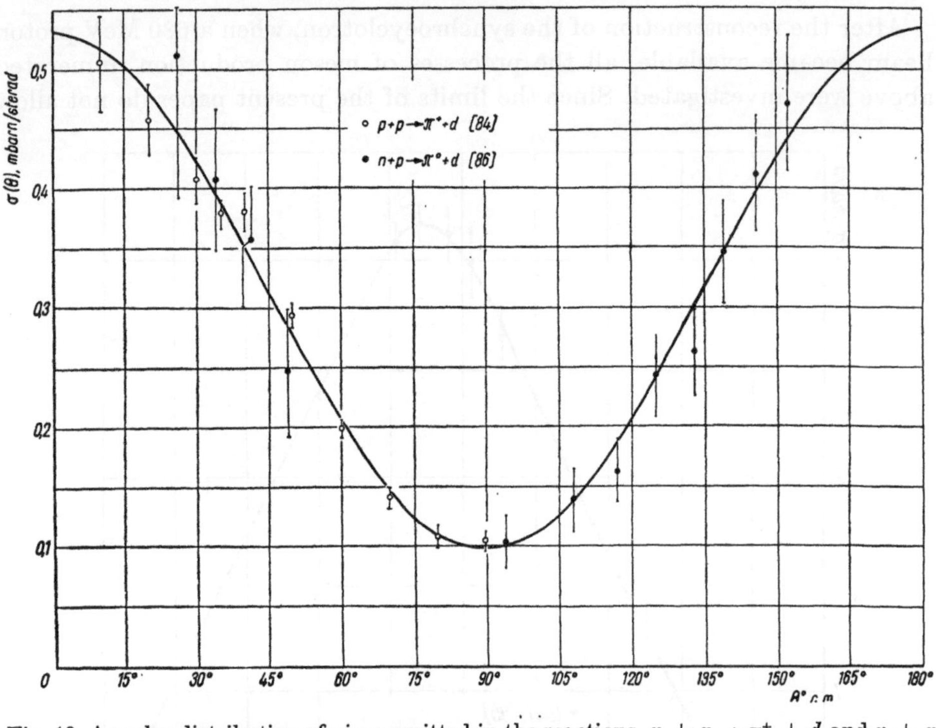

Fig. 19. Angular distribution of pions emitted in the reactions $p + p \rightarrow \pi^+ + d$ and $n + p \rightarrow \pi^0 + d$ for energies of incident protons 650 MeV and neutrons 600 MeV respectively. The data for the second reaction are given in relative units. The identical behaviour of the differential cross-sections for both reactions is seen from the principle of charge independence of (strong) interaction processes and therefore confirms this principle

produced in P-state. This conclusion was made also on the basis of an experimental study (86) at neutron energy of about 600 MeV of the angular distribution of π^0-mesons produced in the reaction $n + p \rightarrow \pi^0 + d$ (fig. 19) which from an isotopic invariance standpoint is identical with the reaction $p + p \rightarrow \pi^+ + d$. Together with the two particle reaction taking place in two proton collision the reaction $p + p \rightarrow \pi^+ + n + p$ with protons of 660 MeV (87) (fig. 20) was also investigated.

The production of π^0-mesons on protons and deuterons has been investigated in the energy region up to 680 MeV when bombarding them by protons (88—91) and neutrons (83, 92).

The ratio of the cross-sections $\sigma^{\pi^0}_{np}/\sigma^{\pi^0}_{pp}$ at 630 MeV is about two (88, 90) — a value which charge independence predicts for the case when the meson-nucleon system has isotopic spin $^3/_2$ (93) in the final state.

The dependence upon energy of the cross-sections of π^0-meson production by protons on protons and deuterons was investigated using both internal (89, 96) and extracted (90, 91) proton beams. In fig. 21 the most accurate measurements are given, in which it was possible to establish that the dependence upon energy of the total π^0-meson production cross-section in $(p-p)$-collisions (89) is quite different from that which had been suggested earlier.

Angular distribution of π^0-mesons produced in nucleon-nucleon collisions has been studied at energies 445, 500, 585 MeV (89) and 660 MeV (89, 90, 94). The angular distribution which is anystropic at

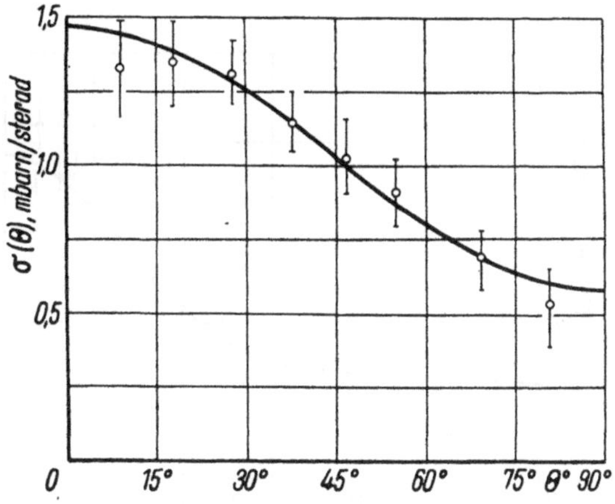

Fig. 20. Angular distribution of π^+-mesons from reaction $p + p \rightarrow \pi^+ + p + n$ at 657 MeV proton energy

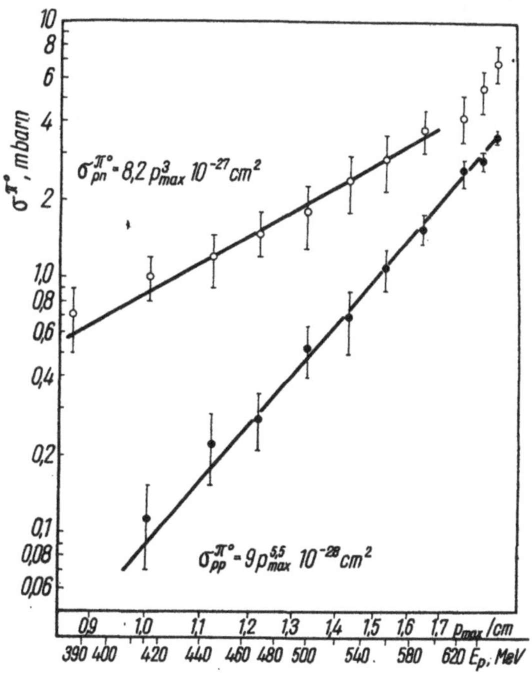

Fig. 21. The energy dependence of total cross-sections of π^0-meson production in $(p + p)$ and $(p + n)$-collisions. E_p is the proton energy, p_{max} is the maximum momentum of π^0-mesons in the centre of mass system in units mc. As can be seen $\sigma^{\pi^0}_{pp}$ increases with proton energy considerably quicker than $\sigma^{\pi^0}_{pn}$, a fact which is due to the „forbidiness" of π^0-meson production in $(p + p)$-collisions near the threshold

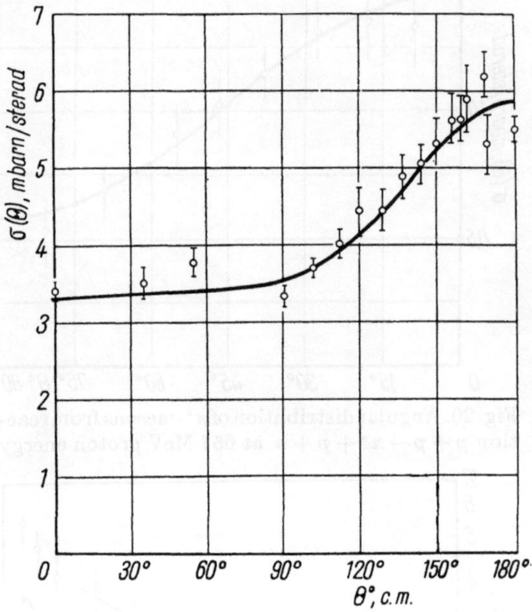

Fig. 22.
Angular distribution of γ-quanta from decay of π^0-mesons produced in carbon by 660 MeV protons. Θ is the effective angle in the centre of mass system colliding nucleons. The decrease of γ-quanta emission at small angle is due to the fact that nucleus is a „thick" absorber both for π^0-mesons and for incident nucleons

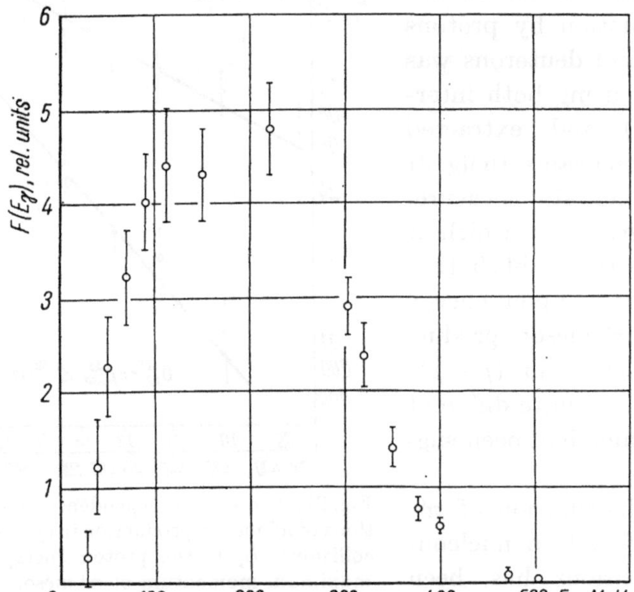

Fig. 23. The energy spectrum of γ-quanta at observation angle of 0° from decay of π^0-mesons produced in ($p-p$)-collisions at 660 MeV proton energy. This spectrum characterizes both the angular distribution and the π^0-meson spectrum. Data cannot be obtained in direct way due to the very small lifetime ($\tau < 10^{-15}$ sec) of the π^0-meson

low energies was found to become practically isotropic when the energy increases up to 660 MeV. This is probably connected with the strong interaction in the final state between the π^0-meson and one of the nucleons.

While investigating the angular distribution of γ-quanta from the decay of π^0-mesons produced on nuclei there was found (88, 90, 94) an interesting asymmetry phenomenon (fig. 22) connected with the fact that the meson and proton mean free paths in nuclear matter are small in comparison with

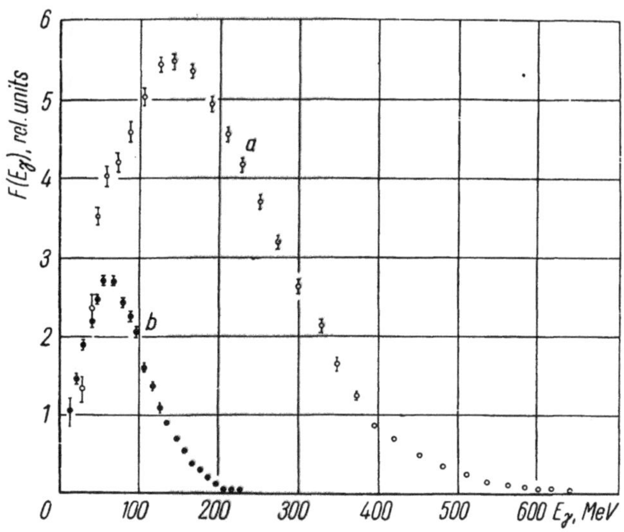

Fig. 24. The energy spectrum of γ-quanta from decay of π^0-mesons produced by 660 MeV protons on carbon nuclei: a) Observation angle 0°, b) Observation angle 180°. The spectrum end of about 600 MeV is connected by those head-on collisions corresponding to events, where one of two γ-quanta from π^0-meson decay practically gets all the energy

the nuclear dimensions. Data on meson absorption in nuclear matter have been obtained also on the basis of measurements of the π^0-meson yields from different nuclei bombarded by protons (88) and neutrons (80, 83).

Energy Spectra of Particles emitted in inelastic Collisions of two Nucleons

Spectra of the particles, emitted in inelastic nucleon collisions with nucleon and nuclei have been investigated using different experimental methods: the method of the analysis with the help of a pair magnetic spectrometer of spectra of γ-quanta from decay of π^0-mesons produced in proton-proton (95) (fig. 23) and proton-nucleus (96, 97) collisions (fig. 24), the method of magnetic analysis of secondary protons and neutrons emitted in proton-proton collisions (98) (fig. 25), the method of magnetic analysis of charged mesons from $(p-p)$ collisions (99, 94) (fig. 26) and from collisions of protons with nuclei (101, 102) (fig. 27) and at last the photoemulsion method (102) (fig. 28) for the analysis of charged meson spectra, produced in $(n-p)$ collisions.

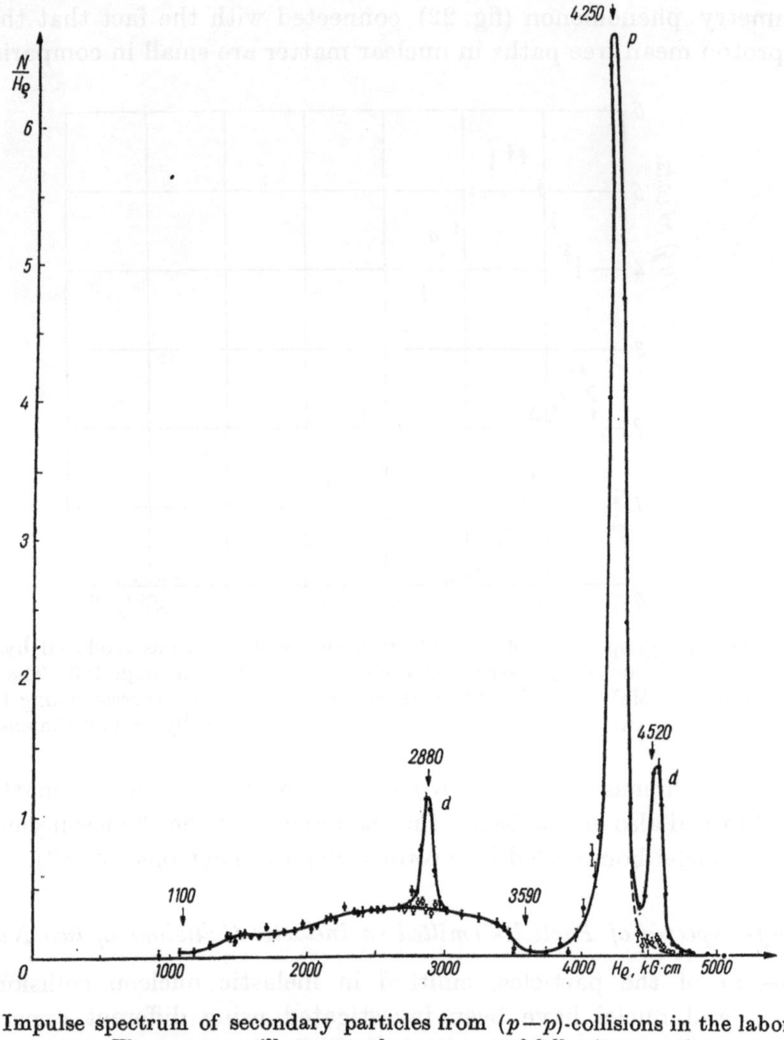

Fig. 25. Impulse spectrum of secondary particles from $(p-p)$-collisions in the laboratory system. The spectrum illustrates the presence of following reactions:

a) elastic $(p-p)$-scattering (proton peak at 4260 kilo gauss × cm,

b) the reaction $p + p \rightarrow \pi^+ + d$ leading to the emission of two deuteron groups with momenta 2880 and 4520 kilo gauss × cm,

c) the reactions $p + p \rightarrow \pi^+ + p + n$ and $p + p \rightarrow \pi^0 + p + \bar{p}$, to which the continuous proton spectrum corresponds

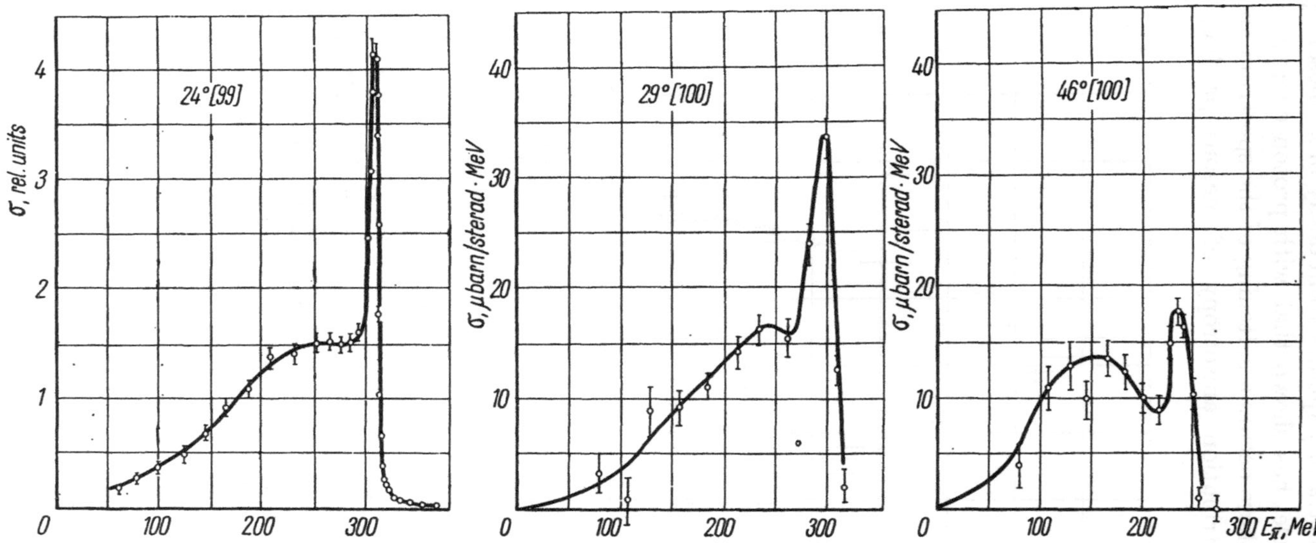

Fig. 26. Energy spectrum of π^+-mesons from the reactions $p + p \rightarrow \pi^+ + p + n$ (1) and $p + p \rightarrow \pi^+ + d$ (2) at angles 24°, 29° and 46° in the laboratory system (the energy of incident protons is 660 MeV). The peak in the spectrum is due to the two particle reaction (2)

The analysis of spectra of γ-quanta from decay of π^0-mesons produced by protons on light nuclei has shown that with proton energy increase from 470 up to 660 MeV there occurs a softening of the spectrum and a decrease of the angular distribution anisotropy of mesons produced in $(n-p)$

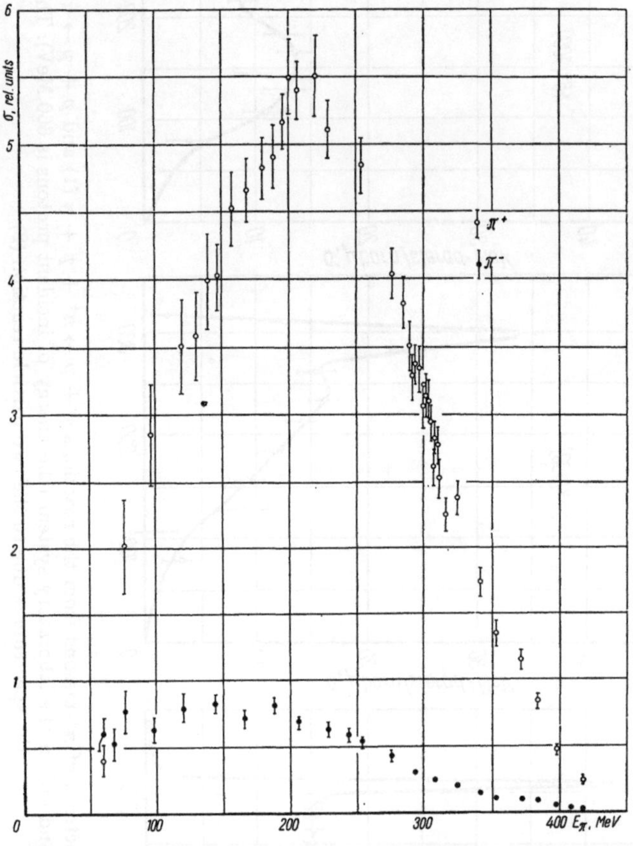

Fig. 27. Energy distribution of π^+ and π^--mesons produced by 660 MeV protons on carbon nuclei at 24° (100). The observed small number of π^--mesons is due to the fact that π^+-mesons may be produced in the nucleus by $(p-p)$ and $(p-n)$-collisions whereas π^--mesons may be produced only in $(p-n)$-collisions which are comparatively ineffective in producing charged mesons

collisions (96). The cause of this is, probably, the appearance of a strong π^0-meson interaction with one of the nucleons. Using the same methods for the first time data were obtained on the angular and energy distribution of π^0-mesons produced in proton collisions with hydrogen (95), the proton energy being 660 MeV.

As a result of the analysis of π^+-meson spectra, produced in the reaction $p + p \rightarrow \pi^+ + p + n$ at proton energies 556 and 657 MeV it was shown that in this energy region the matrix element is proportional to the meson momentum (99).

An interesting result was obtained (102) by measuring the fluxes of charged (102) and neutral (104) mesons emitted from different nuclei when bombarding them with protons. It was found that the ratio of the number

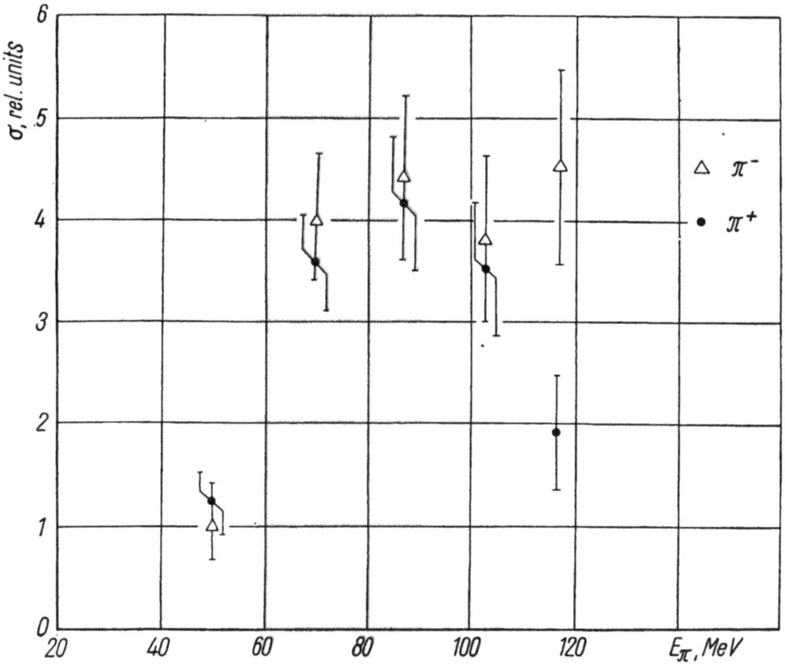

Fig. 28. Spectrum of π^+ and π^--mesons produced by 660 MeV neutrons on hydrogen. The angle of observation is 90°. The data show the equality (within the error) of π^+ and π^--meson yields in agreement with the principle of charge symmetry

of charged mesons to the number of π^0-mesons, produced by protons on nuclei with isotopic spin 0 is equal to two — a fact which confirms once more the hypothesis of charge independence of nuclear forces.

In conclusion it can be noted that such a systematical investigation of the production of neutral and charged π-mesons remains unique up to now.

Production of Mesons by Mesons

It has been known long ago that mesons may be produced in π-meson collisions with nucleons. However, the processes of meson production by mesons up to now had been studied only in an energy region considerably exceeding the meson production threshold.

The investigation of these processes near the threshold has been undertaken at the Laboratory of Nuclear Problems where meson production by mesons has been investigated on liquid hydrogen (105) using scintillation counters and on complex nuclei (106) using photoplates. In fig. 29 the

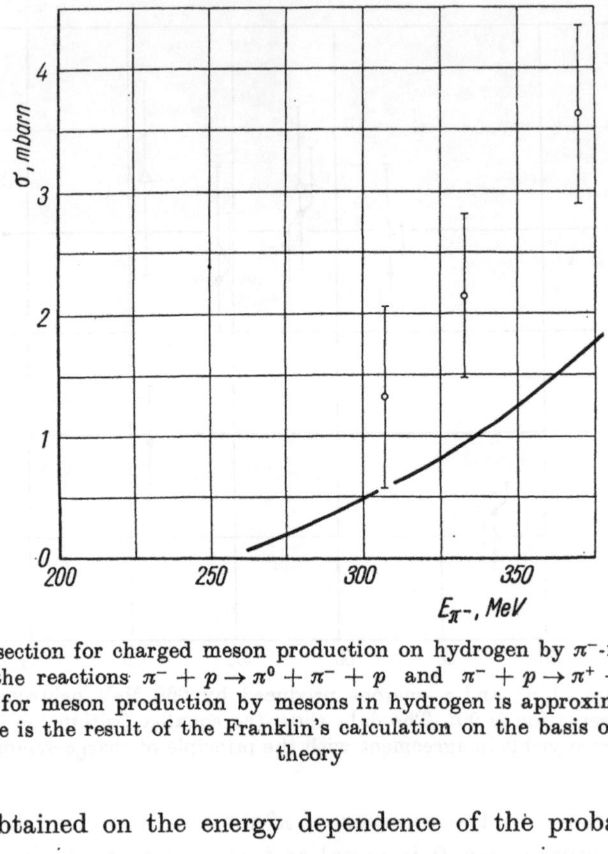

Fig. 29. Cross-section for charged meson production on hydrogen by π^--mesons near the threshold in the reactions $\pi^- + p \rightarrow \pi^0 + \pi^- + p$ and $\pi^- + p \rightarrow \pi^+ + \pi^- + n$ (105). The threshold for meson production by mesons in hydrogen is approximately 170 MeV. The solid curve is the result of the Franklin's calculation on the basis of the Chew-Low theory

first data obtained on the energy dependence of the probability of the reactions

$$\pi^- + p \rightarrow \pi^0 + \pi^- + p \text{ and } \pi^- + p \rightarrow \pi^+ + \pi^- + n$$

are given. All the attempts to calculate the probability of these processes using meson theory failed. The comparison of the experimental data with curve calculated by Franklin on the basis of Chew and Low theory (Phys. Rev. *105*, 401. 1957) which is given in the same figure may be of some interest from a methodical point of view. Chew and Low theory gives better agreement with experiments than other calculations.

Conclusion

Summarizing the results of all the investigations on the interaction of mesons with nucleons and complex nuclei as well as on the meson production by nucleons on nucleons and complex nuclei one may state that they give a full enough phenomenological picture of the meson-nucleon interaction in the energy region up to 0,7 BeV. The main characteristics of this interaction which must be taken into account by the future theory are as follows:

a) meson-nucleon interactions are charge independent,

b) the meson-nucleon interaction is particularly strong in the state with isotopic spin and angular momentum $= {}^3/_2$.

μ-Mesons

It is well known that μ-mesons are not particles strongly interacting with nuclei in the sense that their interaction with nucleons is considerable less than the π-meson-nucleon interaction. However, the question as to whether relatively weak nuclear forces due to virtual μ-meson pair emission ($\mu^{\pm}$, μ^0) exist is open. An attempt to detect (107) an hypothetical neutral μ^0-meson showed that this particle, if it exists, is emitted from nuclei bombarded by 670 MeV protons with a probability less or of the order of 10^{-4} of π-meson emission probability. Therefore the contribution of a pair interaction ($\mu^{\pm}$, μ^0) to nuclear forces is insignificant.

Great interest in the problem of μ-meson decay was renewed after the discovery in the USA of parity non-conservation in weak interactions. In the Soviet Union the asymmetry of the electron emitted in μ-meson decay (in the process $\mu^+ \rightarrow e^+ + \nu + \tilde{\nu}$) with respect to the direction of μ-meson emission in the process $\pi^+ \rightarrow \mu^+ + \nu$ has been investigated (108) using photoemulsion method. The experiments confirmed the conclusion about parity non-conservation and, moreover, gave the form of the electron angular distribution with good accuracy.

In connection with the theory of the longitudinal neutrino, suggested by L. Landau (109) according to which the mass of a neutrino is equal to zero and the neutrino is always polarized in the direction (or against the direction) of its motion, the comparison of experimental data with this theory is of great interest. In this connection an experiment (110) was set up in which the dependence on electron energy of the asymmetry of the electrons emitted in μ^+-meson decay in photoplates was determined. The measurements showed that the asymmetry increases for the energetic part of the electron spectrum: the sign of the effect is in agreement with the predictions of the „longitudinal neutrino" theory. At present other experiments on this fundamental problem are being conducted.

Strange Particles

At the Laboratory of Nuclear Problems Λ^0-particle production was investigated in 1954. On the basis of theoretical considerations, both in the USSR (111) and abroad, the associated production of heavy mesons and hyperons had been predicted a long time ago. Later the law of "associated production" was confirmed by direct experiments conducted at the cosmotron in the USA. The reaction nucleon + nucleon $\rightarrow$ nucleon + Λ^0 contradicts the hypothesis of associated production of Λ^0-particle and K-meson while the reaction nucleon + nucleon $\rightarrow \Lambda^0 + \Lambda^0$ is not in contradiction with such hypothesis. An attempt was made to observe Λ^0-particle production when bombarding the internal target of the synchrocyclotron by 680 MeV protons (112) that is protons with an energy quite sufficient even for the production of two Λ^0-particles. The experiment showed that Λ^0-particles are not produced under these conditions. This appeared to be not only an additional and strong argument in favour of the assumption on associated production of Λ^0-particle and heavy meson but made it also possible to draw a conclusion about the magnitude of the isotopic spin of the K^0-meson taking part in the Λ^0-particle nucleon interaction. It follows from the investigation that K^0-meson and its antiparticle $\tilde{K}^0$ are not identical.

Interaction of High Energy Particles with Complex Nuclei

Apriori one cannot answer the question whether π-meson and nucleon interaction with nuclei is the sum of pair interactions with separate nucleons of nuclei. The main interest of the investigations of the interaction of these particles with nuclei is that they may yield unexpected information on nuclear forces which could not be obtained from experiments on nucleon-nucleon interaction.

We have already mentioned several investigations in which the nucleon and π-meson interactions with complex nuclei were studied. Since this paper is restricted by rather narrow limits we have no possibility to discuss even in short all the investigations in which this interaction was studied in different aspects. However, in order to give an idea of the problems investigated we present some of the publications with the complete title in the list of references (113—121).

Methods of Investigation and Apparatus

The wide program of nuclear investigations performed by means of the synchrocyclotron with particles and γ-quanta of high energies required the

Fig. 30. The measuring pavilion with experimental equipment arranged for work with various particle beams

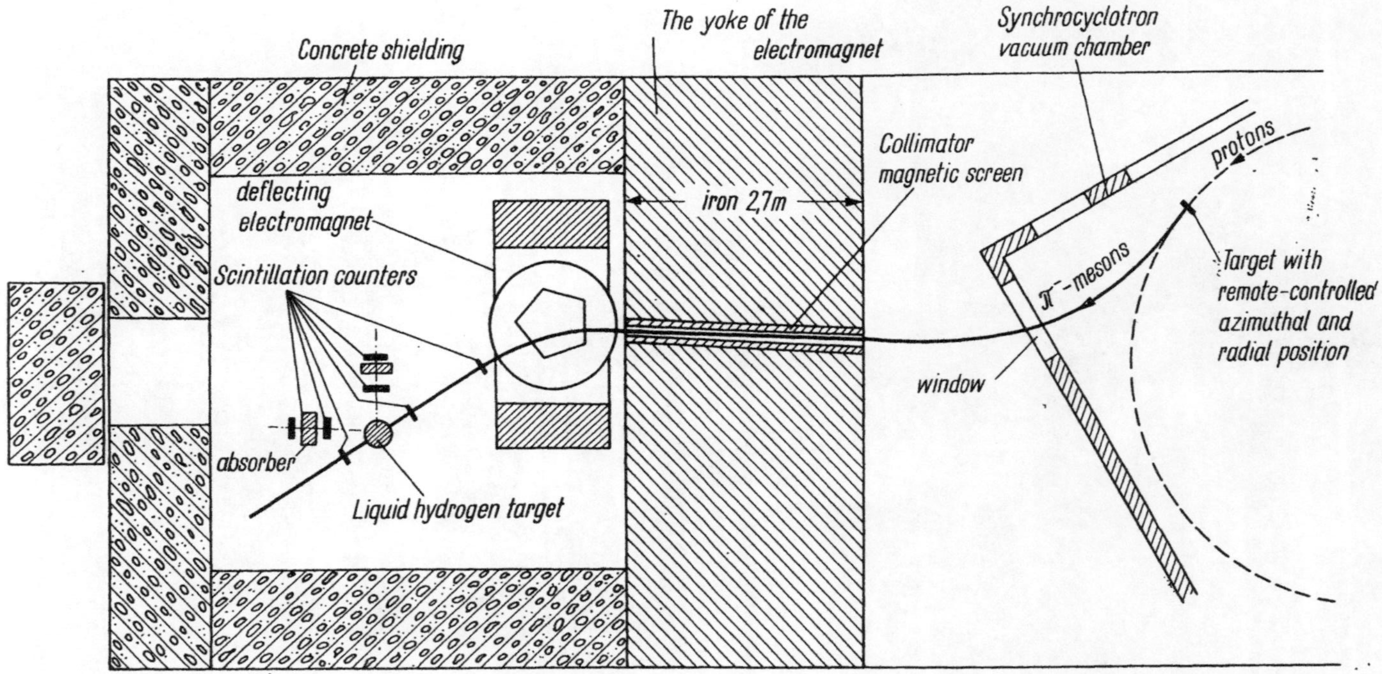

Fig. 31. A typical experimental arrangement for the study of high energy charged π-meson scattering by protons

development of new methods for recording particles and the construction of complicated experimental devices.

The limits of this paper do not allow to make even a short review of all the different experimental apparatus constructed and used in the invistigations set up by many groups of physicists in the course of eight years at the synchrocyclotron. However, when touching upon the whole problem it

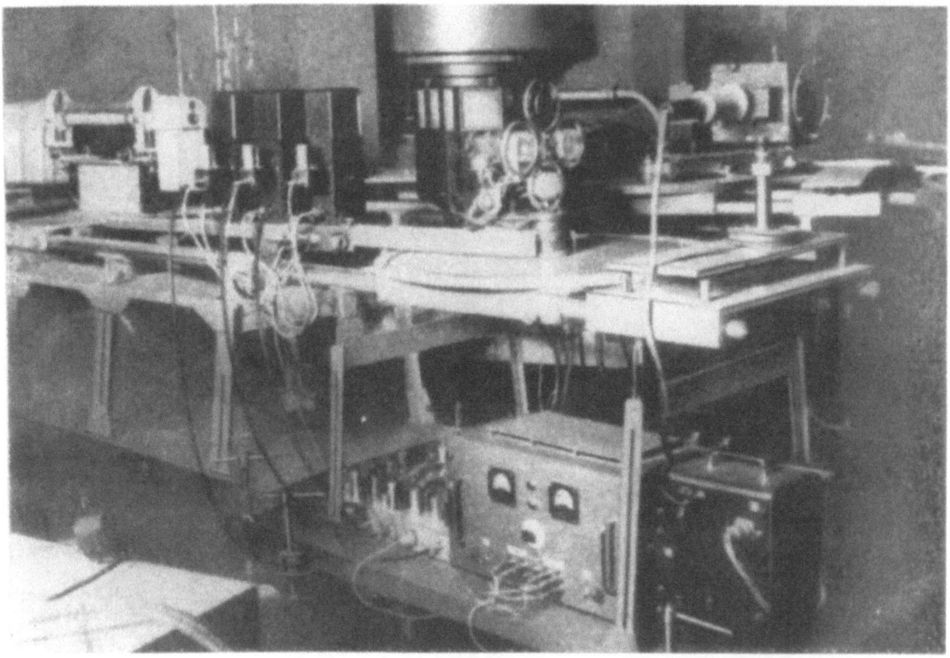

Fig. 32. Experimental set-up for the investigation of high energy proton scattering by protons and neutrons. The system of scintillation counter telescopes detects both secondary particles

should be noted that from the point of view of the difficulty and originality of solutions of scientific-technical problems, as well as from the point of view of the volume, the working out of the research and design works, necessary methods and the construction of the apparatus for experiments with high energy particles belong to one of the most complicated problems which our physicists and engineers had to solve during last eight years.

In order to give a general idea of this aspect of the investigations and to illustrate the scale of the relative development work, we will dwell in short upon some of the devices which were used in the experiments. With this aim we add to the article a few photographs of a number of typical experimental devices. The number of such devices is about fifty. In the course of the experiments they are usually placed on the path of particle beams, in the main measuring pavilion (fig. 30), as well as in the meson and polarization

Fig. 33. A hodoscope system of counters used in experiments on high energy π-meson scattering on hydrogen. The solid angle system is two sterads. The resolving time is $2 \cdot 10^{-6}$ sec. Pulsed power supply makes it possible to use conventional Geiger counters for recording high energy particles under conditions of high background present in the accelerator building

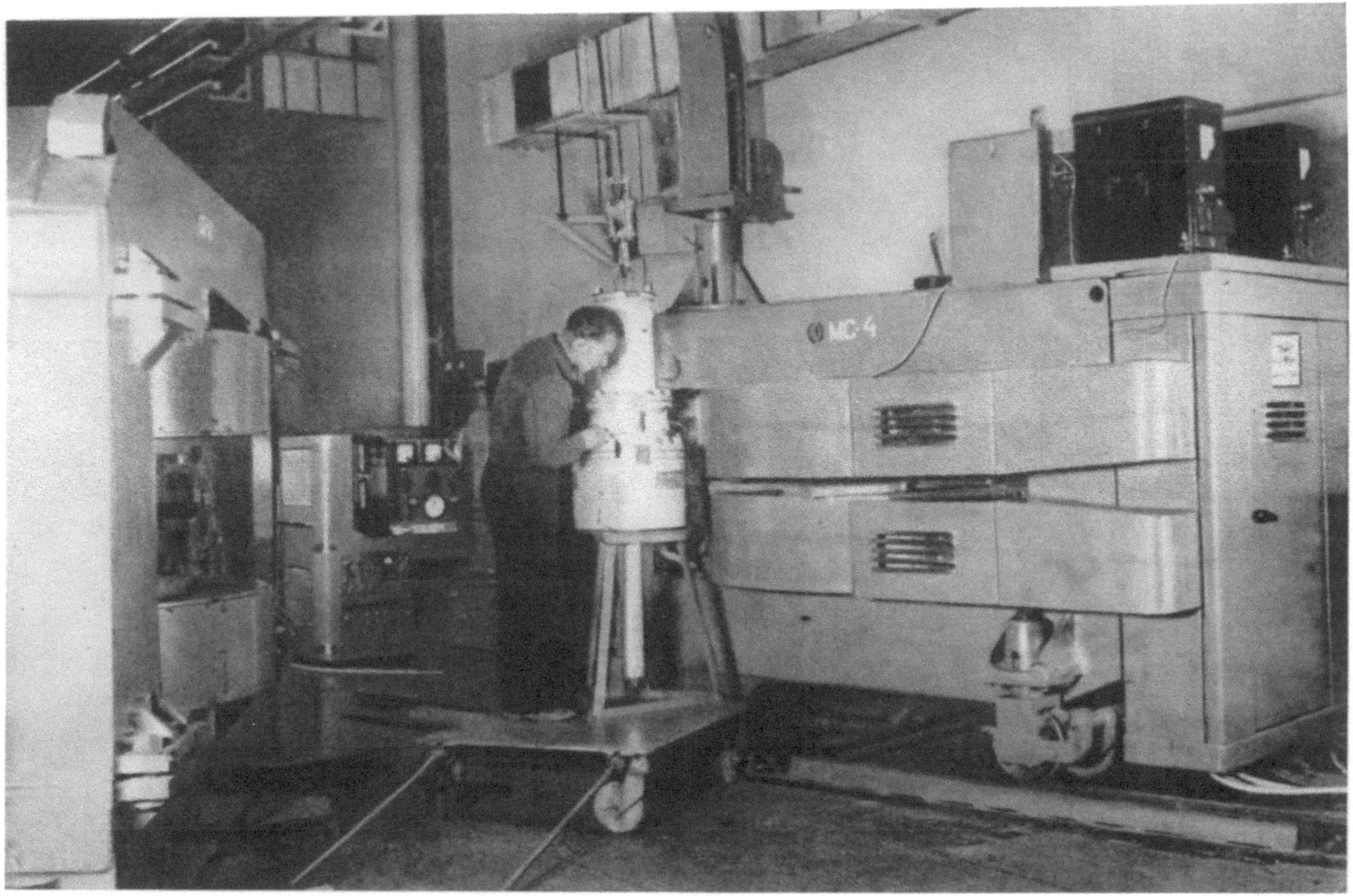

Fig. 34. High pressure diffusion chamber with magnetic field is used for studying the high energy proton interaction with protons and helium nuclei. The chamber (300 mm in diameter) is filled with hydrogen or helium at a pressure of 20 atm. A 16000 oersted pulsed magnetic field is available in the working volume of the chamber

laboratories. Their control is remotely done from a special hall where the radiation level is perfectly tolerable and in which all the outlet recording instruments are placed.

In investigations of the scattering of nucleons and π-mesons by nucleons and nuclei telescopes of crystal and liquid scintillation counters, in combination with photomultipliers switched in coincidence with resolving times up to 10^{-8} sec have been used. Great help in obtaining the scintillation

Fig. 35. A 4-liter bubble chamber with propane filling, meant for effective recording of processes of interaction and decay of elementary particles

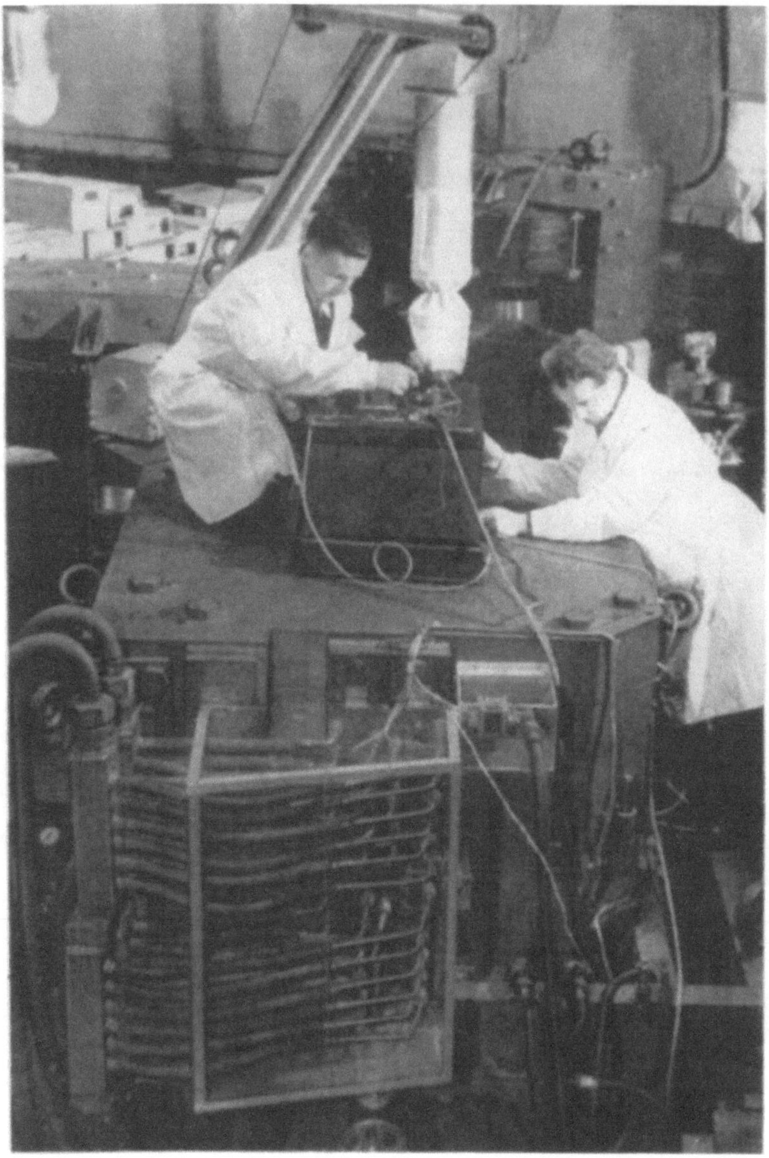

Fig. 36. The magnetic cloudchamber is applied in the investigation of high energy π-meson interaction with atomic nuclei. The effective diameter of the chamber is 400 mm, and the effective height is 100 mm. The intensity of the impulsed magnetic field is 13500 oersted. A deflecting electromagnet can be seen from behind

Fig. 37. General view of the arrangement for investigating inelastic collisions of polarized protons with protons. The secondary particles are detected by telescopes consisting of large scintillation counters. The hydrogen-deuterium target (in the figure it is surrounded by a casing) keeps the liquid in the working cylinder 3,9 cm in diameter for 50 hours. To make background measurements the liquid hydrogen or deuterium can be removed from the working cylinder into a special container from which it can be pumped backward without losses

Fig. 38. The magnetic spectrometer used for investigating the reaction $n + p \rightarrow \pi^0 + d$. The set-up simultaneously records both products of the reaction: the deuteron is detected by proportional counters placed in such a way to yield an analysis of its momentum in a magnetic field, π^0-meson is recorded by counting γ-quanta with scintillation counters telescopes

crystals of different sizes for these counters was rendered by the Institute of Crystallography of the USSR Academy of Sciences. In fig. 31 a typical arrangement of the experiments on scattering of mesons by protons and in fig. 32 a photography of a typical device for investigating nucleon-nucleon scattering are given. In experiments of similar purpose effective hodoscopic systems of several hundred Geiger counters with impulse supply (fig. 33),

Fig. 39. One liter liquid hydrogen bubble chamber with control system (Thermal Laboratory of the USSR Academy of Sciences). The chamber is used for investigating processes of interaction of high energy particles with hydrogen

Čerenkov counters, diffusion chambers with diameter up to 40 cm, filled with hydrogen and helium at a pressure of about 20 atm (fig. 34), bubble chambers with propane (fig. 35) and a cloudchamber with a diameter of 40 cm placed in the magnetic field (fig. 36) have been used and are used.

Experimental devices with the help of which particle scattering or meson production are studied make often use of special targets filled with liquid hydrogen or deuterium. In fig. 37 a general view of one of these liquid hydrogen-deuterium targets is given.

A considerable number of the investigations on charged and neutral π-meson production in nucleon collisions with nucleons and nuclei was made by using many-channel magnetic spectrometers in combination with gas or scintillation counters (fig. 38). The electromagnets of these spectrometers have polar pieces with linear dimensions of the order of a meter and give fields of 15000—17000 oersted in gaps of the order of 10—15 cm. The weight of the magnets is 35—50 tons. The electromagnet-solenoids for the cloud and diffusion chambers have approximately the same parameters. All

Fig. 40. Three section magnetic lens for the focusing of high energy charged particles beams. The focus distance of the lens for protons with a momentum (1200 $\frac{\text{MeV}}{\text{C}}$ is 2,5 m, the aperture is 80 mm, the magnetic field gradient is 2000 oersted/cm

these electromagnets were constructed at the Research Institute of Electrophysical Apparatus.

In experiments on charged π-meson production in neutron-proton collisions a bubble chamber filled with liquid hydrogen (fig. 39) is being used.

Elementary acts of interaction were also observed using nuclear emulsions and emulsion chambers. When scanning the emulsions the microscopes are equipped with special devices manufactured at the laboratory.

Fig. 41. Twenty-channel amplifier for signals from scintillation counters. The amplification of each channel is 40 decibel, the bond with 140 megacycles

Fig. 42. The reprojector for the determination of the location in space of charged particle tracks obtained (in stereophotographs of cloud, diffusion and bubble chamber events). The diameter of the screen is 400 mm

In a number of experiments the necessity arises of a preliminary beam focusing. This is done by electromagnetic quadrupole lenses, one of which is shown in fig. 40.

Complicated conditions in performing physical experiments with the synchrocyclotron required the construction of a great deal of specific electronic apparatus: fast amplifiers, multichannel analyzers etc. The necessity arose of working out also stereocameras, reprojectors and stereocomparators, which are used in connection with cloudchamber, diffusion chamber and bubble chamber work. The photos of some devices are given in fig. 41, 42. A considerable amount of nuclear investigations with the six-meter synchrocyclotron has been also made using radiochemical methods.

CONCLUSION

Shortly reviewing the physical investigations made by means of the six-meter synchrocyclotron of the Laboratory of Nuclear Problems since its starting up, it can be said that these investigations have made a valuable contribution to the physics of high energy particles. They extended our knowledge in this field of modern physics and at the same time helped to put forward new important problems and to work out the methods for their further investigation.

To a considerable extent thanks to these investigations a young science, high energy physics, has occupied its stable place among all other sciences in our country.

The work at the six-meter synchrocyclotron has been a good school for a considerable number of scientists from several research institutes as well as for engineers and designers of different specialities who assimilated new modern technique.

During the last year (1956—1957) because of the establishment of the Joint Institute for Nuclear Research a considerable number of scientists from other countries — members of the Institute — have joined and, therefore, the laboratories of the Institute became an international nuclear research centre.

One of the tasks of which is to prepare physicists for all the twelve Countries, Members of the Institute.

The starting up of the giant synchrofasotron of 10 BeV at the High Energy Laboratory of the Joint Institute for Nuclear Research has opened new great additional perspectives for the further development of nuclear physics.

LIST OF REFERENCES

1. D. V. Efremov, M. G. Mescheryakov, A. L. Mints, V. P. Dzhelepov, P. P. Ivanov, *V. S. Katyshev*, E. G. Komar, I. F. Malyshev, N. A. Monoszon, I. H. Nevyashski, B. I. Polyakov, A. V. Chestnoi, Atomnaya Energiya, N 4, 5 (1956).
2. M. G. Mescheryakov, A. V. Chestnoi, V. P. Dzhelopov, *V. S. Katyshev*, A. A. Kropin, B. I. Zamolodchikov, V. P. Dmitrievski, E. L. Grigorjev, *A. A. Reut*, A. L. Savenkov, A. G. Vahrameev, T. N. Tomilina, V. V. Batyunya, A. E. Ignatenko, S. N. Yurov, RINP, 1955.
3. A. L. Mints, I. H. Nevyazhski, B. I. Polyakov, CERN Symposium I, 419 (1956).
4. B. I. Zamolodchikov, Report of the Laboratory of Nuclear Problems*) of the Joint Institute for Nuclear Researches, 1957.
5. V. P. Dzhelepov, V. P. Dmitrievski, *V. S. Katyshev*, M. S. Kozodaev, M. G. Mescheryakov, K. I. Tarakanov, A. V. Chestnoi, Atomnaya Energiya, N4, 13 (1956).
6. A. E. Ignatenko, V. V. Krivitski, A. I. Muhin, B. M. Pontecorvo, *A. A. Reut*, K. I. Tarakanov, Atomnaya Energiya, N5, 2 (1956).
7. V. P. Dmitrievski, V. I. Danilov. Yu. N. Denisov, N. L. Zaplatin, *V. S. Katyshev*, A. A. Kropin, A. V. Chestnoi, Pribory i Tehnika Eksper. N 1, 11 (1957).
8. A. A. Kropin, RINP JINR, 1957.
9. V. I. Danilov, V. P. Dmitrievski, A. V. Chestnoi, Pribory i Tehnika. Eksper. N3, 9 (1956).
10. M. G. Mescheryakov, Session of the USSR Academiy of Sciences on Peaceful uses of Atomic Energy (plenary meeting), 1955, 39. A. P. Vinogradov, Session of the USSR Academy of Sciences on Peaceful uses of Atomic Energy (plenary meeting) 1955, 60.
11. N. P. Bogachev, RINP, 1950.
12. R. I. Gerasimova, I. I. Gurevich, K. N. Muhin, RINP, 1950.
13. N. P. Bogachev, RINP, 1951.
14. N. A. Perfilov, N. S. Ivanova, RINP, 1950; JETP 29, 551 (1955). N. A. Perfilov, O. V. Lozhkin, V. P. Shamov, JETP 28, 656 (1955).
15. G. E. Belovitski, L. V. Suhov, T. A. Romanova, J. M. Frank, RINP, 1950; JETP 28, 729 (1955).
16. N. A. Perfilov et al. Proceedings of the Radium Institute, USSR Academy of Sciences, 7, 3—98 (1956).
17. V. P. Dzhelepov, Yu. M. Kazarinov, B. V. Gavrilovski, B. M. Golovin, RINP, 1950.
18. M. G. Mescheryakov, E. L. Grigorjev, N. P. Bogachev, L. M. Soroko, RINP, 1950.
19. B. V. Gavrilovski, RINP, 1950.
20. M. S. Kozodaev, A. A. Markov, RINP, 1950.
21. V. P. Dzhelepov, B. M. Golovin, Yu. M. Kazarinov, RINP, 1950 and 1951.
22. B. V. Kurchatov, V. N. Mehedov, M. Ya. Kuznetsova, L. N. Kurchatova, RINP, 1950. B. V. Kurchatov, V. N. Mehedov, N. I. Borisova, M. Ya. Kuznetsova, L. N. Kurchatova, L. V. Chistyakov, Session of the USSR Academy of Sciences on peaceful uses of atomic energy (section of chemical sciences) 1955, 178.
23. B. T. Geilikman, Proceedings of the Academy of Sciences of the USSR, 1950.
24. *A. A. Reut*, G. I. Selivanov, V. V. Yurjev, RINP, 1950. V. I. Goldanski, V. S. Penjkina, E. Z. Tarumov, RINP, 1951; Dokl. Acad. Nauk SSSR, 101, 1027 (1955); JETP 29, 778 (1955).
25. S. A. Baranov, RINP, 1950.
26. M. G. Mescheryakov, N. P. Bogachev, B. S. Neganov, E. V. Piskarev, RINP, 1951. E. L. Grigorjev, L. M. Soroko, RINP, 1950. A. P. Zhdanov, Z. S. Sokolova, Report of the Radium Institute of the USSR Academy of Sciences, 1950. A. P. Zhdanov, Z. S. Sokolova, Report of the RIAN USSR, 1952.

*) Further: Laboratory of Nuclear Problems of the Joint Institute for Nuclear Researches — LNP JINR.

27. L. P. Solovjeva, JETP, 31,12 (1956). A. P. Zhdanov, F. G. Lepehina, Report of the RIAN USSR, 1951.
28. V. I. Goldanski, A. N. Kovalski, V. S. Penjkina, E. Z. Tarumov, RINP, 1951; Dokl. Akad. Nauk, SSSR, 106, 219 (1956).
29. V. I. Goldanski, A. E. Ignatenko, A. I. Muhin, V. S. Penjkina, V. A. Shkoda-Ulyanov, RINP, 1951. Communicated by V. I. Goldanski at the Gordon conference on nuclear chemistry (USA), June, 1957.
30. Ya. A. Smorodinski, Proceedings of the USSR Academy of Sciences, 1953; Problems sovremennoi fiziki, 7, 7 (1954).
31. M. G. Mescherjakov, N. P. Bogachev, B. S. Neganov, E. V. Piskarev, RINP, 1952; Dokl. Akad. Nauk SSSR, 99, 955 (1954). B. S. Neganov, L. M. Soroko, I. K. Vzorov, Dokl. Acad. Nauk SSSR, 99, 959 (1954). M. G. Mescheryakov, N. P. Bogachev, B. S. Neganov, Izv. Acad. Nauk SSSR, ser. fiz. 19, 548 (1955).
32. N. P. Bogachev, I. K. Vzorov, RINP, 1954; Dokl. Acad. Nauk SSSR 99, 931 (1954).
33. N. P. Bogachev, Dokl. Acad. Nauk SSSR, 108, 806 (1956).
34. S. Ya. Selector, S. Ya. Nikitin, E. G. Bogomolov, E. G. Zombkovski, Izv. Acad. Nauk SSSR, ser. fiz. 19, 561 (1955); Dokl. Acad. Nauk SSSR, 99, 967 (1954).
35. E. G. Bogomolov, E. G. Zombkovski, S. Ya. Nikitin, S. Ya. Selector, CERN Symposium 2, 129 (1956).
36. M. G. Mescheryakov, S. B. Nurushev, G. D. Stoletov JETP (in press).
37. V. P. Dzhelepov, B. M. Golovin, V. I. Satarov, RINP, 1952; Dokl. Acad. Nauk 99, 934 (1954).
38. B. M. Golovin, V. P. Dzhelepov, JETP 31, 194 (1956).
39. V. P. Dzhelepov, V. I. Satarov, B. M. Golovin, RINP, 1951; Dokl. Acad. Nauk SSSR, 104, 717 (1955); JETP 29, 369 (1955).
40. V. P. Dzhelepov, V. I. Moskalev, S. V. Medved, Dokl. Acad. Nauk, SSSR, 104, 380 (1955).
41. V. P. Dzhelepov, Yu. M. Kazarinov, RINP, 1951; Dokl. Acad. Nauk SSSR, 99, 939 (1954).
42. V. P. Dzhelepov, Yu. M. Kazarinov, B. M. Golovin, V. B. Flyagin, V. I. Satarov, Izv. Acad. Nauk SSSR, ser. fiz. 19, 573 (1955). V. P. Dzhelepov, Yu. M. Kazarinov, B. M. Golovin, Yu. N. Simonov, CERN Symposium 2, 115 (1956).
43. Yu. M. Kazarinov, Yu. N. Simonov, JETP 31, 169 (1956).
44. N. S. Amaglobeli, Yu. M. Kazarinov. RLNP JINR, 1957.
45. I. Ya. Pômeranchuk, Dokl. Acad. Nauk SSSR, 78, 349 (1955); JETP, 21, 1113 (1956).
46. V. P. Dzhelepov, Yu. M. Kazarinov, V. B. Flyagin RINP, 1952; Dokl. Acad. Nauk SSSR 100, 655 (1955).
47. M. G. Mescheryakov, E. V. Piskarev, N. P. Bogachev, B. S. Neganov, RINP, 1952. M. G. Mescheryakov, N. P. Bogachev, G. A. Leksin, B. S. Neganov, E. V. Piskarev, CERN Symposium 2, 125 (1956).
48. G. A. Leksin, JETP. 32, 445 (1957).
49. L. S. Azhgirei, I. K. Vzorov, V. P. Zrelov, M. G. Mescheryakov, B. S. Neganov, A. F. Shabudin, JETP (in press).
50. D. I. Blohintsev, JETP (in press).
51. V. M. Sidorov, E. L. Grigorjev, JETP (in press). O. V. Lozhkin, N. A. Perfilov. JETP 31, 913 (1956).
52. V. P. Dzhelepov. V. I. Moskalev, Dokl. Acad. Nauk SSSR, 110, 532 (1956).
53. A. E. Ignatenko, A. I. Muhin, E. B. Ozerov, B. M. Pontecorvo, Dokl. Acad. Nauk 103, 45 (1955).
54. A. E. Ignatenko, A. I. Muhin, E. B. Ozerov, B. M. Pontecorvo, Dokl. Adac. Nauk SSSR 103, 209 (1955).
55. A. E. Ignatenko, A. I. Muhin, E. B. Ozerov, B. M. Pontecorvo, JETP 30, 7 (1956).
56. A. I. Muhin, E. B. Ozerov, B. M. Pontecorvo, JETP 31, 371 (1956).
57. A. I. Muhin, B. M. Pontecorvo JETP 31, 550 (1956).

58. M. S. Kozodaev, R. N. Sulyaev, A. J. Filipov, Yu. A. Scherbakov, Dokl. Acad. Nauk SSSR, 107, 236 (1956).
59. L. S. Duljkova, T. A. Romanova, I. B. Sokolova, L. V. Suhov, K. D. Tolstov, M. G. Shafranova, Dokl. Acad. Nauk SSSR, 107, 43 (1956). L. S. Duljkova, I. B. Sokolova, M. G. Shafranova, Dokl. Acad. Nauk SSSR, 111, 992 (1956).
60. V. G. Zinov, S. M. Korenchenko, JETP (in press).
61. E. L. Grigorjev, N. A. Mitin, JETP 31, 37 (1956).
62. V. G. Zinov, S. M. Korenchenko (to be published).
63. N. A. Mitin, E. L. Grigorjev JETP 32, 440 (1957).
64. I. E. Tamm, Yu. A. Goljfand, V. Ya. Fainberg, JETP 26, 649 (1954).
65. A. E. Ignatenko, A. I. Muhin, E. B. Ozerov, B. M. Pontecorvo, Dokl. Acad. Nauk SSSR, 103, 395 (1955).
66. A. E. Ignatenko, A. I. Muhin, E. B. Ozerov, B. M. Pontecorvo, JETP 31, 545 (1956).
67. V. G. Ivanov, V. T. Osipenkov, N. I. Petrov, V. A. Rusakov, JETP 31, 1097 (1956).
68. N. A. Mitin, E. L. Grigorjev, Dokl. Acad. Nauk SSSR, 103, 219 (1955).
69. L. P. Kudrin, B. A. Nikoljski, Dokl. Acad. Nauk SSSR, 111, 795 (1956).
70. V. P. Dzhelepov. V. G. Ivanov, M. S. Kozodaev, V. T. Ossipenkov, N. I. Petrov, V. A. Rusakov, JETP 31, 923 (1956).
71. M. S. Kozodaev, R. M. Sulyaev, A. I. Filippov, Yu. A. Scherbakov, JETP 31, 701 (1956).
72. M. S. Kozodaev, R. M. Sulyaev, A. I. Filippov, Yu. A. Scherbakov, JETP (in press).
73. B. M. Pontecorvo, G. I. Selivanov, RINP, 1952; Dokl. Acad. Nauk SSSR, 102, 495 (1955).
74. M. S. Kozodaev, A. A. Tyapkin, R. A. Vanetsian, RINP, 1953.
75. B. M. Pontecorvo, G. I. Selivanov, V. A. Shukov, RINP, 1953.
76. L. M. Soroko, RINP, 1953.
77. M. G. Mescheryakov, B. S. Neganov, N. P. Bogachev, V. M. Sidorov, RINP, 1953; Dokl. Acad. Nauk SSSR, 100, 673 (1955).
78. M. S. Kozodaev, A. A. Tyapkin, A. A. Markov, Yu. D. Bayukov, RINP, 1953.
79. M. S. Kozodaev, A. A. Tyapkin, Yu. D. Bayukov, A. A. Markov, Yu. D. Prokoshkin, Izv. Acad. Nauk, ser. fiz. 19, 589 (1955).
80. B. M. Pontecorvo, G. I. Selivanov, RINP, 1952; Dokl. Acad. Nauk SSSR, 102, 253 (1955).
81. B. M. Pontecorvo, G. I. Selivanov, RINP, 1951.
82. B. M. Pontecorvo, Ya. A. Smorodinski, Proceedings of the Second conference on high energy physics, Dubna, 1953.
83. V. P. Dzhelepov, K. O. Oganesyan, V. B. Flyagin, JETP 32, 678 (1957).
84. M. G. Mescheryakov, B. S. Neganov, Dokl. Acad. Nauk SSSR, 100, 677 (1955).
85. B. S. Neganov (to be published).
86. V. P. Dzhelepov, V. S. Kiselev, K. O. Oganesyan, V. B. Flyagin (to be published).
87. B. S. Neganov, O. V. Savchenko, JETP (in press).
88. A. A. Tyapkin, M. S. Kozodaev, Yu. D. Prokoshkin, Dokl. Acad. Nauk SSSR, 100, 689 (1955).
89. Yu. D. Prokoshkin, A. A. Tyapkin, JETP 32, 750 (1957).
90. Yu. D. Balashov, V. A. Zhukov, B. M. Pontecorvo, G. I. Selivanov, RINP 1955.
91. L. M. Soroko JETP 30, 296 (1956).
92. V. P. Dzhelepov, K. O. Oganesyan, V. B. Flyagin, JETP 29, 886 (1955).
93. L. I. Lapidus, RINP, 1955.
94. Yu. D. Prokoshkin, A. A. Tyapkin, RINP, 1955. Yu. D. Prokoshkin, CERN Symposium 2, 385 (1956).
95. Yu. D. Bayukov, A. A. Tyapkin, JETP 32, 953 (1957).
96. Yu. D. Bayukov, M. S. Kozodaev, A. A. Tyapkin JETP 32, 667 (1957).
97. Yu. D. Bayukov, A. N. Sinaev, A. A. Tyapkin, JETP 32, 385 (1957).

98. M. G. Mescheryakov, B. S. Neganov, I. K. Vzorov, V. P. Zrelov, A. F. Shabudin, Dokl. Akad. Nauk SSSR, 190, 499 (1956).
99. M. G. Mescheryakov, V. P. Zrelov, B. S. Neganov, I. K. Vzorov, A. F. Shabudin, JETP 31, 45 (1956).
100. A. G. Meshkovski, Yu. S. Pligin. Ya. A. Shalamov, V. A. Shebanov, JETP 31, 560 (1956).
101. M. G. Mescheryakov, I. K. Vzorov, V. P. Zrelov, B. S. Neganov, A. F. Shabudin, JETP 31, 55 (1956).
102. A. G. Meshkovski, Yu. S. Pligin, Ya. Ya. Shalamov, V. A. Shebanov, JETP 31, 987 (1956); JETP (in press).
103. V. I. Sidorov, JETP 28, 727 (1955).
104. Yu. D. Prokoshkin, A. A. Tyapkin, JETP (in press).
105. V. G. Zinov, S. M. Korenchenko (to be published).
106. V. V. Krivitski, *A. A. Reut*, Dokl. Acad. Nauk SSSR, 112, 232 (1957).
107. A. N. Novikov, B. M. Pontecorvo, G. I. Selivanov, JETP 29, 889 (1955).
108. I. I. Gurevich, A. P. Mishakova, B. A. Nikolski, L. V. Surkova, JETP (in press).
109. L. D. Landau, JETP 26, 649 (1957).
110. A. O. Vaisenberg, V. A. Smirnitski, JETP (in press).
111. B. M. Pontecorvo, RINP, 1951 and 1953; JETP 29, 140 (1955).
112. M. P. Balandin, B. D. Balashov, V. A. Zhukov, B. M. Pontecorvo, G. I. Selivanov, JETP 29, 265 (1955).
113. V. P. Shamov, O. V. Lozhkin, Asymmetry of fragment ranges in disintegration of heavy nuclei by ultrafast particles, JETP 29, 286 (1955). N. A. Perfilov, V. I. Ostroumov, On the processes of star production and fission of W and Bi nuclei bombarded by 460 MeV protons, Dokl. Acad. Nauk SSSR, 103, 227 (1955). V. I. Ostroumov, Fission of U nuclei by fast protons, Dokl. Acad. Nauk SSSR, 103, 409 (1955). N. S. Ivanova, N. A. Perfilov, V. P. Shamov, Fission of U nuclei by 460 MeV protons, Dokl. Acad. Nauk SSSR, 103, 573 (1955).
114. E. L. Grigorjev, Detection of polarization of fast particle beams by means of photoemulsiontechnique, JETP 28, 761 (1955).
115. M. G. Mescheryakov, S. B. Nurushev, G. D. Stoletov, Polarization of 660 MeV protons scattered by nuclei, JETP 31, 361 (1956).
116. *V. V. Alpers*, L. M. Barkov, R. I. Gerasimova, I. I. Gurevich, K. N. Muhin, B. A. Nikolski, E. L. Toporkova, Production of slow pions in photographic emulsion nuclei by 460 MeV protons and neutrons of 400 MeV effective energy, JETP 30, 1025 (1956). *V. V. Alpers*, L. M. Barkov, R. I. Gerasimova, I. I. Gurevich, A. P. Mishakova, K. N. Muhin, B. A. Nikolski, Production of slow pions in photographic emulsion nuclei by 660 MeV protons, JETP, 30, 1034 (1956).
117. *A. A. Reut*, S. M. Korenchenko, V. V. Yurjev, B. M. Pontecorvo, Detection of H^4 nuclei among the products of Carbon fission induced by 300 MeV protons, Dokl. Acad. Nauk SSSR, 102, 723 (1955).
118. E. L. Grigorjev, L. P. Solovjeva, Nuclear fission induced by 660 MeV protons in photographic emulsion, JETP 31, 932 (1956).
119. V. S. Kiselev, V. B. Flyagin, Energy distributions of neutrons emitted from Beryllium bombarded by 680 MeV protons, JETP 32, 957 (1957). V. P. Dzhelepov, Yu. M. Kazarinov, B. M. Golovin, V. B. Flyagin, Emission of high energy neutrons induced by 480 MeV protons, RINP, 1953.
120. Yu. P. Kumekin, Polarization in quasielastic $(p-p)$ scattering of 653 MeV protons on deuterons, JETP (in press).
121. B. M. Pontecorvo, A. E. Ignatenko, Measurement of the total cross-sections for interactions of 300 MeV neutrons with nuclei by means of star detector, RINP, 1951. V. I. Moskalev, B. V. Gavrilovski, Total cross section for interactions of 650 MeV protons with nuclei, Dokl. Acad. Nauk SSSR, 110, 972 (1956).

Bei Fragen zur Produktsicherheit wenden Sie sich bitte an:
If you have any questions regarding product safety,
please contact:

Walter de Gruyter GmbH
Genthiner Straße 13
10785 Berlin
productsafety@degruyterbrill.com